Advances in Geographic Information Science

Series editors
Shivanand Balram, Burnaby, Canada
Suzana Dragicevic, Burnaby, Canada

More information about this series at http://www.springer.com/series/7712

Liliana Perez • Eun-Kyeong Kim • Raja Sengupta
Editors

Agent-Based Models and Complexity Science in the Age of Geospatial Big Data

Selected Papers from a workshop on Agent-Based Models and Complexity Science (GIScience 2016)

Springer

Editors
Liliana Perez
Laboratoire de Géosimulation Environnementale
Department of Geography
Université de Montréal
Montreal, QC, Canada

Eun-Kyeong Kim
Department of Geography
Pennsylvania State University
University Park, PA, USA

Raja Sengupta
Department of Geography
School of Environment
McGill University
Montreal, QC, Canada

ISSN 1867-2434 ISSN 1867-2442 (electronic)
Advances in Geographic Information Science
ISBN 978-3-319-65992-3 ISBN 978-3-319-65993-0 (eBook)
DOI 10.1007/978-3-319-65993-0

Library of Congress Control Number: 2017953010

Printed on acid-free paper

This Springer imprint is published by Springer Nature
The registered company is Springer International Publishing AG
The registered company address is: Gewerbestrasse 11, 6330 Cham, Switzerland

Preface

A broad range of concepts and methodologies from complexity science—including agent-based models (ABMs), cellular automata (CA), and network theory, among others—have contributed to a better understanding of spatiotemporal dynamics of complex geographic patterns and processes. Particularly, ABMs have become ubiquitous in GIScience and a number of related application domains, prompting some ABM researchers to propose the YAAWN syndrome. Along with ABMs, much more scaling relations have been found through geospatial big data analytics. However, such convergence is not unidirectional. Many statistical (social) physicists have done research on human mobility, urban dynamics, and landscape dynamics, which have traditionally been the domain of geographers and environmental scientists. Recent advances in computational technologies such as big data, cloud computing and CyberGIS platforms, and sensor networks (i.e., the Internet of things) provide new opportunities and raise new challenges for ABM and complexity theory research within GIScience. With growing accessibility to rich, big data sources and increased computing power, geographers can simulate dynamic geographic phenomena in a more realistic fashion and test theories and models using empirical data. Despite of the utility of complexity theories, adopting those methodologies properly to the geographic domain is an ongoing research issue. Challenges include parameterizing the complex models with volumes of georeferenced data being generated, scaling the model applications to realistic simulations over broader geographic extents, exploring the problems in their deployment across large networks to take advantage of increased computational power, and validating their output using real-time data, as well as measuring the impact of the simulation on knowledge, information, and decision-making both locally and globally via the World Wide Web.

In September of 2016, the Ninth International Conference on Geographic Information Science (GIScience) was held in Montreal, Canada, and brought together approximately 300 participants from around the world from academia, industry, and government organizations to discuss and advance the state of the art in geographic information science. Within the context of GIScience, we held a workshop named

"Rethinking the ABCs: Agent-Based Models and Complexity Science in the Age of Big Data, CyberGIS, and Sensor Networks." The scope of this workshop was to explore novel complexity science approaches to dynamic geographic phenomena and their applications, addressing challenges and enriching research methodologies in geography in a big data era. The 1-day workshop brought together experts on complexity science and social networks in order to discuss novel complexity science approaches to dynamic geographic phenomena and their applications, addressing challenges and enriching research methodologies in geography in a big data era. We had nine lightning talks and nine presentations, corresponding to four short and five full peer-reviewed papers. We wrapped up the workshop with a very interesting discussion about the future of agent-based models. As a result of a very productive workshop, it was decided to publish the major findings as a book within the Springer GIScience series. Seven selected papers from the workshop, which reflect the advances on ABM development and implementation, as well as the opportunities that big data and network theory could bring for a better understanding of complex systems, have been included in this book.

The research covered by the collection of papers in this volume offers the reader a possibility to encounter diverse applications of ABMs fully implemented and tested, through the first three chapters, followed by a fourth chapter that presents an ABM to identify human migration pathways. Finally, the last three chapters explore the possibilities of using big data and social networks to parameterize ABMs and discover the complexities of movement, migration, and urban patterns.

Chapter 1 by Cenek and Franklin describes an ABM system for the management of stocks and stakeholders of Alaska's Salmon Fisheries. It uses 35 years of sonar data to parameterize and calibrate the stock information, combined with interviews of fishermen, in order to validate the model. Chapter 2 by Bitterman and Bennett presents and discusses the potential of using ABMs to explore resilience concepts in an agricultural land use system. The authors suggest a novel approach in terms of exploring the concept of resilience as an adaptive behavior within a complex system. Chapter 3 by Taylor and Dragicevic presents a very interesting work applying the invariant-variant approach for validating an insect dispersal ABM. As evaluating the performance of agent-based models is notoriously difficult, the presented work makes an interesting and valuable contribution. Chapter 4 by Arnoux et al. offers a novel approach to identify migration pathways due to armed conflicts by proposing an ABM to simulate human decision-making to migrate from conflict areas. Chapter 5 by Sengupta et al. offers a novel perspective about the use of big data in order to extract movement rules to parameterize an ABM of animal mobility. Chapter 6 by Liu et al. presents a very interesting work using spatial network visualizations and IRA database to understand migrations across the United States, revealing the urban hierarchy by investigating the directional network structure of US cities created based on the US migration patterns. Last but not least, Chapter 7 by Koylu presents some innovative ideas related to the leveraging of social media to understand human interactions at a level that was difficult or impossible to do using traditional interaction data.

Finally, we would like to express our appreciation to all contributing authors for their excellent work. Their participation made our workshop a major success and made this book possible. We also thank the program committee and additional reviewers for reviewing and sharing their experience. We thank the many people who made GIScience 2016 possible: the steering committee for their support and the local organizing committee. Last but not least, we would like to warmly thank our colleagues and families for supporting us.

Program Committee

Clio Andris (Pennsylvania State University), David Bennett (University of Iowa), Christopher Bone (University of Victoria), Suzana Dragicevic (Simon Fraser University), Bin Jiang (University of Gävle), Alan M. MacEachren (Pennsylvania State University), Mir Abolfazl Mostafavi (Universite Laval), Atsushi Nara (San Diego State University), David O'Sullivan (University of California, Berkeley), and Taha Yasseri (Oxford Internet Institute, University of Oxford).

Lighting Talks

David Bennett – *Dependent on Which Path? Complexity and Agent-Based Modeling*
Daniel G. Brown – *Combining Spatial-Temporal Data with Behavioral Models Helps Us Better Understand Spatial Process*
Suzana Dragicevic – *A Perspective on Voxel-Based Geographic Automata*
Bin Jiang – *The Third Definition of Fractal*
Eun-Kyeong Kim – *Burstiness Measure as a New Exploratory Spatio-Temporal Data Analysis Statistic*
David O'Sullivan – *Simple Simulation Models as a Complexity 'Pattern Language'*
Mir Abolfazl Mostafavi – *CAMUSS: The State of the Art in Cellular Automata*
Raja Sengupta – *What Can We Learn from Big Data? Behavioural Rule Extraction from Animal Movement Databases*
Clio Andris – *System Resilience and Collapse as a Function of the Informed Agent*

Montreal, QC, Canada — Liliana Perez
University Park, PA, USA — Eun-Kyeong Kim
Montreal, QC, Canada — Raja Sengupta

Contents

About the Editors

Liliana Perez is the director of the Laboratory of Environmental Geosimulation (LEDGE) and an assistant professor in the Department of Geography at the University of Montreal. Liliana is interested in advancing GIScience methods applied to ecology by developing modeling approaches to simulate ecological complexities in order to understand their behavior and dynamics as well as to use them as a starting point to begin planning and preparing management strategies in the face of climate change. She has developed and implemented a series of simulation tools focusing on forestry, landscape ecology, biodiversity, and climate change.

Eun-Kyeong Kim is a Ph.D. candidate in the GeoVISTA Center in the Department of Geography at Pennsylvania State University. Eun-Kyeong has been developing spatiotemporal data analysis methodologies by adopting approaches from statistical physics and complexity science and is interested in geospatial big data visualization with advanced technologies. She has served as a graduate researcher for an NSF-sponsored big data education project, and she is a coauthor of an online textbook on big data analytics.

Raja Sengupta is associate professor in the Department of Geography and School of Environment at McGill University. Dr. Sengupta is interested in research on both artificial life and software agents and applying GIScience to environmental management issues and water resources management. He was an editorial board member for the journal *Transactions in GIS* (2011–2016) and is currently an editorial board member for *Water International*.

Developing High Fidelity, Data Driven, Verified Agent Based Models of Coupled Socio-Ecological Systems of Alaska Fisheries

Martin Cenek and Maxwell Franklin

Abstract Alaska salmon fisheries are a source of commercial revenue, renewable subsistence resource, cultural identity, and recreational destination for Alaskans, native populations, and out of state eco-tourists alike. We constructed a high fidelity, adaptable, data-driven agent based model that generalizes the socio-ecological dynamics of Kenai River, Alaska. Interactions among the model's agents can be altered to study the impact of fishing regulation changes or salmon run-timing dynamics. Agents are driven by stochastic principles derived from 35 years of integrated data including salmon runs, municipality management reports, and Alaska Department of Fish and Game management reports. Longitudinal and seasonal correlations between the model's simulation outputs and the reported system measurements are used to validate the model.

Keywords Socio-ecological dynamics • Fisheries • Agent-based model • Data driven

1 Introduction

Alaska salmon are a source of commercial revenue, renewable subsistence resource, and cultural identity for the Alaska Native populations as well as a source of recreation for Alaskan residents and out of state visitors. The commercial salmon harvest alone exceeds 400 million a year in economic revenue [2]. Building a high-fidelity, verified model of the coupled social and ecological systems, salmon and society, is necessary to understand the dynamics of individual systems and the mutual interplay between the fisheries and society. The model is intended to be used as a decision support tool for effective governance and resource management. Fishery managers can test the outcomes of a proposed policy using scenario-based testing and simulations, or study the impact of changes to the social and

M. Cenek (✉) • M. Franklin
University of Alaska Anchorage, Anchorage, AK 99508, USA
e-mail: mcenek@uaa.alaska.edu; mefranklin@alaska.edu

L. Perez et al. (eds.), *Agent-Based Models and Complexity Science in the Age of Geospatial Big Data*, Advances in Geographic Information Science,
DOI 10.1007/978-3-319-65993-0_1

ecological drivers on the coupled systems dynamics. We constructed an ABM that accurately generalizes both the salmon runs and the annual harvest by all major stakeholder groups. The model design includes fusion of biophysical and social data sources, translating the measured system dynamics into agents and environments, measuring the outputs of the constructed model across multiple inter- and intra-system dynamics, and finally calculating correlations between the reported system dynamics and the measured model outcomes.

The coupled socio-ecological systems do not have clearly defined geo-physical or social boundaries, nor are they defined by a set of descriptive inter- and intra-system interaction dynamics or common units of dynamics measurements. First, we collected, fused, interpolated and inferred system dynamics from multiple data sources. The resulting data-sets establish biophysical and behavioral observations that were used to model individual agent behaviors and interactions with other agents and the environment. The scope of data-collection was limited to support investigation of multiple hypotheses. Next, we defined both longitudinal and seasonal correlation metrics to measure the model's performance against the collected data-sets. Finally, we developed a statistically based computational framework that uses the recorded agent behaviors as input and analyzes them to produce a state-space transition network that captures the agents' prototypical behaviors exhibited during the model execution. Currently, we are adapting the Geometry of Behavioral Spaces Framework to measure the model's sensitivity to the parameter changes that drive the ABM. Since the analysis only concerns the agent behaviors, the scenario based experimentation can be implemented as changes to multiple model parameters that drive the agent behavior.

We built the ABM to support the analysis of several hypotheses that include understanding of how altered salmon runs may affect personal-use fisheries, how effective the various fishermen groups are to manage the seasonal salmon escapement goal, and how a policy change for commercial fisheries may affect sport fisheries. Although the model currently uses data-sets that describe the Kenai River fisheries, the ABM is parametrized and data driven so it can be easily altered to model most of the fisheries given the availability of data. To support the model's primary goal to study the nature of coupled socio-ecological system behaviors, the model's simulation area is not spatially explicit, but instead the watershed of interest is generated at random according to the overall watershed characteristics. This cardinal notion of space is used to simply limit the agent interaction opportunities without spatially over-fitting model behavior.

The model's agents represent two species of salmon and four types of fishermen. An average of approximately 3.7 million Sockeye salmon have been returning to the Kenai River in recent years to spawn, requiring the maximum sustainable harvest of approximately 3.0 million salmon, and allowing for approximately $700,000$ salmon to escape for spawning. Chinook salmon on the other hand are coveted trophy fish with returning numbers in tens of thousands. The fishermen groups represented in the model include the personal-use sport fishermen in the upper reaches of the watershed and the dipnetters limited to netting the fish from the river mouth up to approximately river mile 4 [2]. The commercial fishery agents represent the drift gillnet fleet operating in the ocean waters of the Upper Cook Inlet (UCI) and the set gillnet fleet that are permitted to anchor nets from the North and South beaches adjacent to the Kenai River mouth. The local interactions of stakeholder agents with each other and the landscape give rise to system wide complex dynamics. These patterns include the total number of salmon allowed to 'escape' and successfully spawn (also called the *escapement goal*) [10], the annual harvest counts for various fishermen groups, and more complex metrics such as catch per unit of effort (CPUE). Fishery managers use escapement goals for the chinook and sockeye salmon in choosing whether to issue an emergency opening or closure [2]. The different run sizes and timing dynamics of the chinook and sockeye salmon and their combined escapement goals represent a complex goal that the Alaska Department of Fish and Game (ADFG) managers attempt to meet every season.

Building an accurate model of the coupled systems dynamics has to integrate measured data sources from (1) bio-physical measurements of the salmon runs, (2) the harvest data from all stakeholders involved in the fisheries, and (3) the spatial-temporal coupling between the bio-physical and social dynamics. The ABM design illustrates how we de-coupled highly interconnected systems, disambiguated the collected longitudinal measured dynamics, and inferred the information about the fishermen behavior that is not known, but is included in the reported coupled system dynamics.

2 Determining Socio-Ecological Dynamics

2.1 Reconstructing Salmon Run-Timing Dynamics

Calculating the returning salmon counts and the run-timing dynamics is a mosaic process of compiling, dis-aggregating, and adjusting data from multiple data-sources. The biophysical salmon data is measured from in-river sonar and the genetic sampling of randomly selected harvested salmon in UCI [8]. The measurement of the social system's interaction with the fisheries is reported as the daily Sockeye harvest and Chinook by-catch counts from all stakeholder groups.

The Sockeye salmon run-timing dynamics at Kenai River are measured at a single point by Dual Frequency Identification Sonar (DIDSON); currently, the Sockeye salmon sonar is located at river mile 19 and the Chinook salmon sonar at river mile 14 [13]. The species counts are calculated by processing the sonar video feeds from both river banks using Adaptive Resolution Imaging Sonar method. Additional information about the in-river salmon populations are collected from random sampling by netting [2]. Published sonar counts are not adjusted for the salmon harvested downstream at the mouth of the river and in Upper Cook Inlet. To appropriately seed the returning salmon agents in the model, the run-timing dynamics from the sonar data must be adjusted for the harvested salmon in the ocean and the Kenai River prior to the sonar counter.

To model accurate salmon agents counts, we reconstructed the temporal distribution of the salmon run by taking 35 years of reported sonar counts as the baseline, adding the dipnet harvest, and adding the salmon harvest of the set and drift gillnet fleets. The genetic sampling of randomly selected salmon caught by the drift gillnet, set gillnet, as well as test fisheries at the mouth of Upper Cook Inlet is used to determine how much salmon caught by the off-shore fishermen were returning to the Kenai River watershed instead of the rest of the inlet tributaries [4]. Time frames of these harvests are reported alongside estimated harvest from genetic sampling by gear type.

Salmon runs were grouped into four categories by their overall characteristics of run-timing dynamics. We used the sonar records to categorize the run-timing patterns using feature-scaling to filter daily sockeye counts with values $x' \geq 0.5$ (Eq. 1). The temporally aligned and weekly binned series of filtered sonar data for 35 years were mutually compared. The distributions with high similarity were grouped into the four resulting prototype categories. Averaging the series in each prototype category produced the generalized baseline time-series distribution of the salmon runs with variance margins (Fig. 1).

$$x' = (x - min(x))/(max(x) - min(x)) \tag{1}$$

Using estimates of Kenai salmon commercial harvest from genetic sampling, we calculated the adjustments to the baseline sonar counts for each week of the season from July 1 to August 15 for reported years 2005–2011 [4]. The same proportional adjustments were inferred for the years without genetic sampling conducted. Similar to the commercial fisheries, the seasonal dipnet harvests for the fishing season from July 10 to July 31 were added to the generalized sonar distributions without using genetic stock identification since all salmon harvested were slotted for spawning in the Kenai River watershed.

The reconstructed run-timing dynamics combined the sonar baseline counts, Kenai River proportions of the commercial salmon harvests, and the dipnet harvests (and all dynamics were aligned for temporal lag before addition) to generate the salmon agents entering the simulation. The resulting distribution is scaled up and down to reflect the variance in the overall salmon run size. Each ABM simulation first selects the prototype category, after which the number of salmon agents are randomly generated using the generalized distribution within the variance margins. The in-river salmon escapement is measured by the sonar, where as the salmon escapement is measured in the streams of the upper watershed. Therefore the effective escapement is calculated as the difference of the sonar counts and the salmon harvested by the personal and commercial sport fishermen. By adjusting the run-timing dynamics of the model based on harvest timing, reliance upon explicit spatial parameters of the system is reduced since the stakeholders in the simulation are ensured an opportunity to harvest correctly timed and seeded salmon abundance regardless of spatial attributes in the model relative to the system. This allows for the generalized watershed setting used to simulate the fishery dynamics upon in our model.

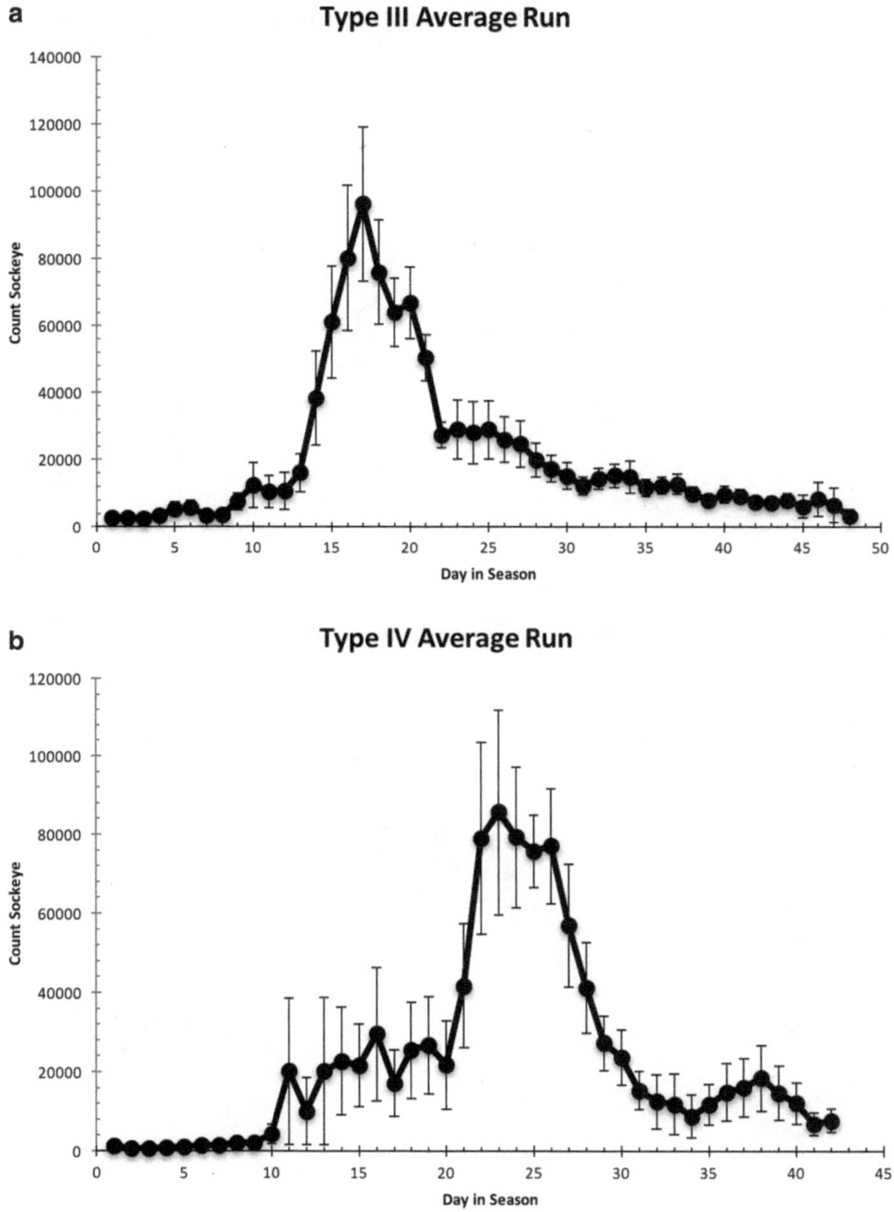

Fig. 1 The feature scaling with $x' \geq 0.5$ classification of each sockeye salmon run-timing dynamics for 35 years of reported sonar data into one of four characteristic classes. Each plot shows the averaged run distribution and the standard error. The run types III, IV and III-IV are named after the peak location in the returning salmon run in week 3, 4, 3–4 of the season respectively. The type III-IV-V has multiple peaks in weeks 3, 4, and 5 of the season

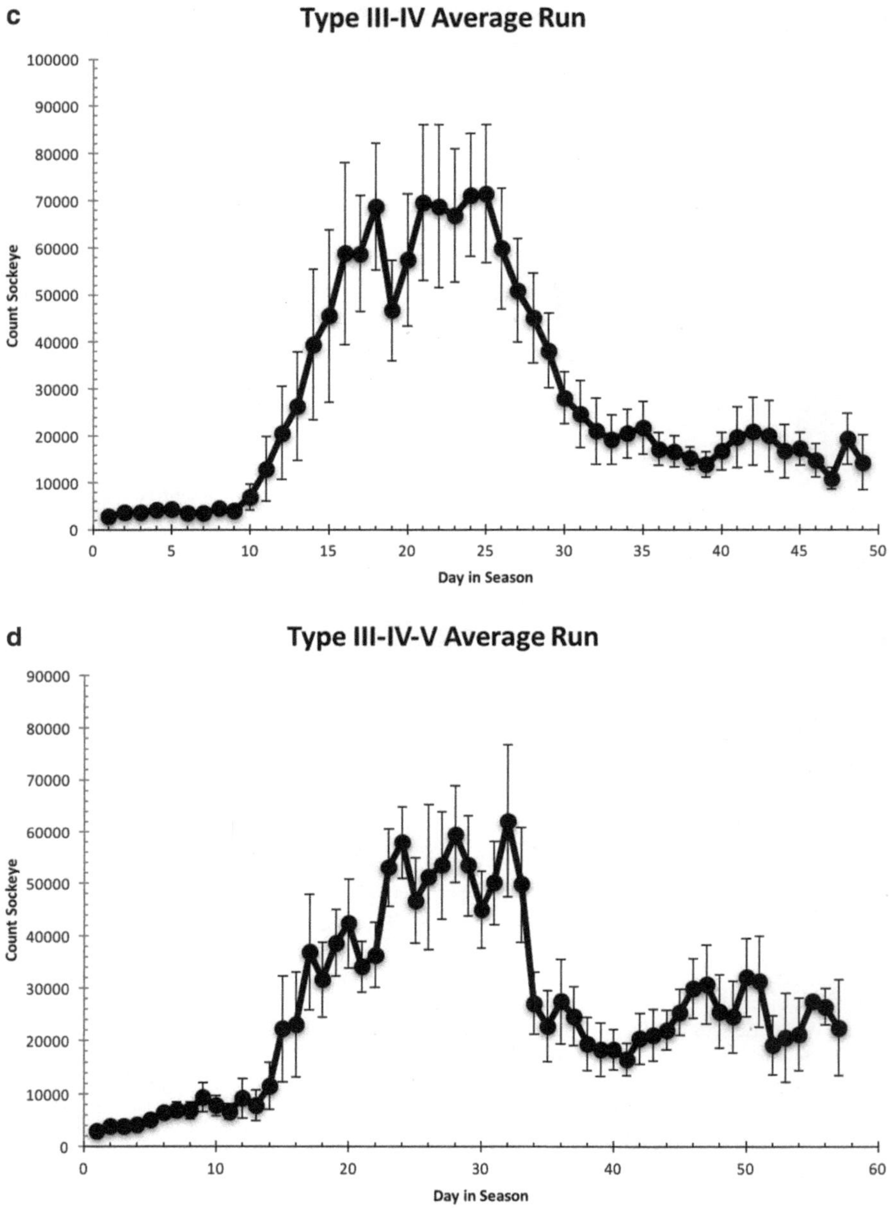

Fig. 1 (continued)

2.2 Coupled Socio-Ecological Systems Dynamics

Salmon harvest represents the coupling between the fishermen effort to catch fish and the salmon run, in addition to other factors such as gear choice, fishing location, and fishing efficiency. The harvest reports are the aggregate socio-ecological metrics used to express the interaction dynamics between the social and ecological systems. Catch Per Unit of Effort (CPUE) is one such measure. CPUE can be used as an index of both stock abundance [10, 11] and stakeholder effort [9]. We decoupled the temporal CPUE distributions into the constituent system dynamics and using the previously reconstructed salmon run-timing distributions, we were able to infer the social behavior that was not previously measured or reported. Building the interaction dynamics of the model's fishermen agents is a reverse inference process. The agents have to have the same behavior as the social behavior inferred from CPUE and when the model's fishermen agents interact with salmon agents, the correlation between the model's CPUE output and the measured CPUE must be high.

The units of CPUE measurements are different for each stakeholder group due to harvests reported at different frequency and fidelity. Dipnet CPUE is reported for each day fished and recorded on household level personal-use dipnet permits. The permits reflect the household size and household seasonal salmon quota. After decoupling dipnet CPUE, designing the dipnet agents' behaviors, and collecting the CPUE data from the model, we calculated the correlation between the reported CPUE by ADFG and the output CPUE from the model simulations. An $R^2 = 93\%$ correlation value for dipnet effort and an $R^2 = 84\%$ correlation for dipnet harvest indicates the model accurately captures the fundamentals of the coupled socio-ecological system dynamics of the dipnet stakeholders.

The CPUE for the drift gillnet fisheries is reported as a cumulative delivery or harvest per fishing season day. Dis-aggregating the cumulative CPUE dynamics is needed so the model can create the correct number of unique drift gillnet agents for each day of the commercial fishing season with appropriate variance of each vessel's probability to catch salmon. Weekly binned, feature scaled and averaged drift-gillnet delivery dynamics in Fig. 2a. were extracted from drift-gillnet CPUE delivery data to generate the number of drift gillnet fishermen agents in the model, while the CPUE harvest was used to calibrate the baseline probability of drift-gillnet agents harvesting the salmon agents. The variance in the CPUE deliveries distributions were used for stochastic generation of the model's drift gillnet agents. We used two different CPUE measures because they reflect unique characteristics of stakeholders that are translated to the agent parameters and they cannot be used synonymously.

Dis-aggregating drift gillnet CPUE delivery data allowed for inference of social behavior of drift gillnet agents from delivery dynamics. Disambiguating and inferring the social system dynamics of set gillnet commercial fishermen used the same analysis principles for the drift gillnet fleet. The set gillnet stakeholders report harvest CPUE and have similar run-timing dynamics.

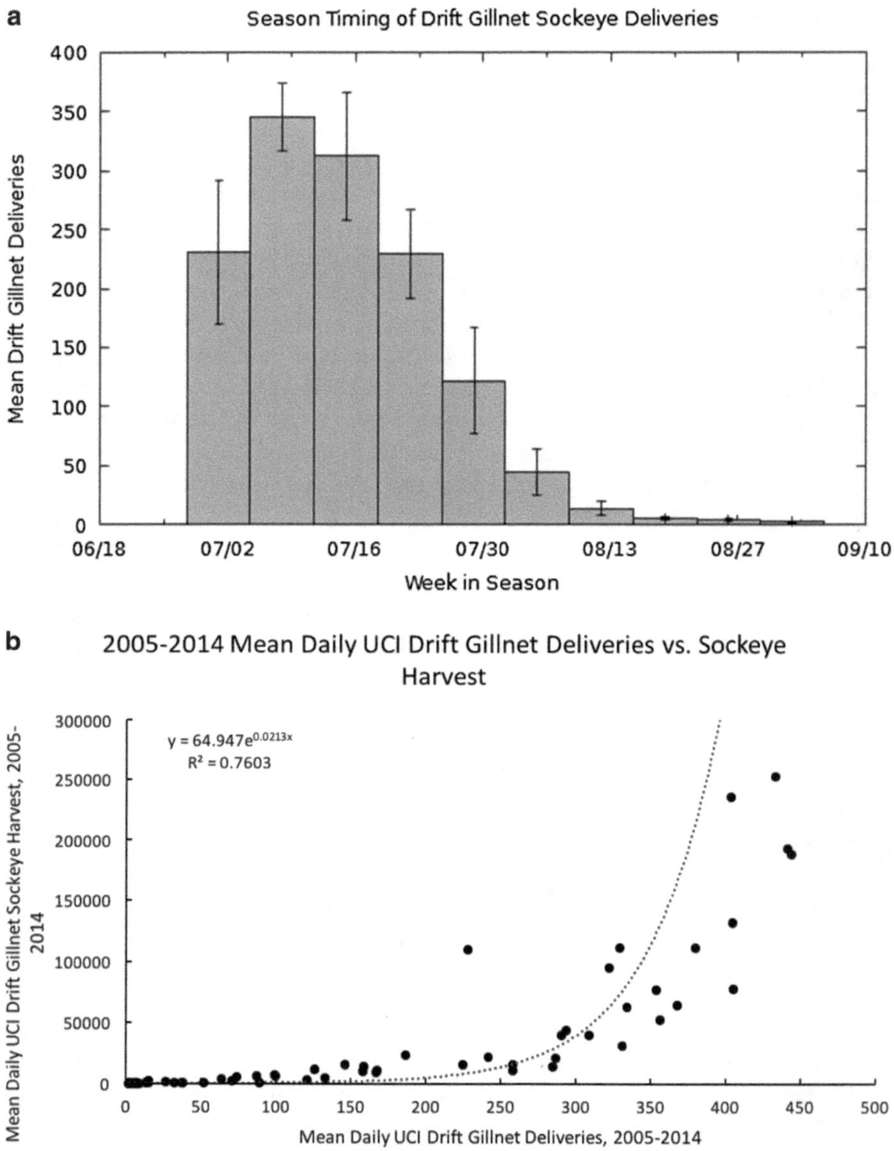

Fig. 2 Historical data on drift gillnet catch-per-season-day were decoupled and binned by week into constituent delivery and harvest by date dynamics. Sockeye deliveries for drift gillnet CPUE with error bars representing standard error are in (**a**). Harvest and deliveries are not synonymous as witnessed in (**b**)

2.3 Reconstructing the Social System's Dynamics

Similar to reconstructing run-timing, social system dynamics have to be assembled from a variety of different data sources. Timing of stakeholder effort is determined by factors including daily deliveries, salmon harvest, annual permits fished, number of days open for fishing, and other services utilized such as parking permits or total revenue generated from the day use permits for dipnet fishing in the Kenai county.

Drift gillnet effort and associated variance is inferred from the feature-scaled, weekly binned, mean delivery data (Fig. 2a). Effort as a function of delivery timing was inferred from the exponential correlation of mean daily drift gillnet deliveries with mean daily drift gillnet harvest across 2005–2014 (Fig. 2b). The exponential regression suggests that deliveries and harvest may be decoupled and the variance in delivery data may capture drift gillnet effort. Due to the run-timing dynamics and effort dynamics for the set gillnet fleet being similar to the drift gillnet fleet, set gillnet effort was modelled with a similar method.

Modelling of commercial fishery stakeholder behavior has previously been conducted with both profit and utility maximization models as well as foraging theory models [14]. The behavior of dipnetters at the Kenai River cannot be modeled by profit and utility models since they participate in personal-use subsistence fisheries of no-profit value. We propose a happiness function to drive dipnet effort and behavior. The Kenai county management reports publish the boat launch counts and revenue, the number of day use permits sold, and the overall daily dipnet revenue [1]. The ADFG published the management reports with the annual dipnet harvests from self-reported harvest counts for 1996–2014 with daily harvests for 2011–2014 [12]. We cross-correlated the daily harvests with the county revenue reports and the annual number of permits fished to model the number of agents fishing for a given period of the season. We found the dipnet effort temporal distribution parallels the observed historical pattern of sockeye run-timing. In particular, the effort increases until approximately mid July, then effort starts to slowly decline. We used this reported behavior to model the temporal behavior of dipnet agents. The dipnet agents fish for a number of days determined by the happiness function (Eq. 2). Happiness is affected by h = current daily harvest, h_h = historical or expected harvest for a day, h_p = previous daily harvest, h_l = mean daily harvest of local stakeholders, m = a motivation factor unique for each agent, H_l = mean happiness of local stakeholders, H_p = previous happiness, and the final term is a decreasing factor with d = number of days fished by the stakeholder. The motivation factor is a constant determined for each agent using a random-normal distribution with a mean of 0.30 and a standard deviation of 0.10.

$$H = ((((h/h_h) + (h/h_p) + (h/h_l))/3) + m)H_l H_p(7/d + 7.75) \qquad (2)$$

The agents in our model follow a fusion of Cartesian and stochastic fishing strategies similar to the theoretical fishing strategies proposed by Cabral et al. [6]. These fishing strategies are representative of either risk-averse or risk-loving fishermen [3, 5, 6]. Agents are required to pass a Cartesian threshold before moving to a new random location during the simulation. This balance of risk-averse and risk-loving behavior avoids the bias of stakeholders in the simulation towards entirely risk-averse or risk-loving behavior. The ABM's Cartesian threshold is determined by an agent decision tree that compares the agent's daily current harvest to the mean daily harvest of local agents, historical or expected daily harvest, and previous daily harvest parameters. If an agent's current daily harvest is below any two of the aforementioned parameters, agents randomly change location to a new fishing ground.

3 The Model

A data driven model must strike a balance between generalization of trends in recorded data and over-fitting the model to describe the recorded data instance. To achieve this balance, the model's behavior uses stochastic sampling with variance of generalized system behaviors to capture the trends in recorded data. To ensure the model did not over-fit the data, the model was constructed using a subset of available data for training while the model validation tested how well the model predicted all recorded data. To validate the model's predictive ability, we intend to compare how well the model predicts dynamics of individual or coupled system dynamics that were not used during the model construction and the dynamics reported by the management agencies.

3.1 Model Validation Using the Coupled Socio-Ecological Systems Dynamics

Model validation primarily uses the coupled system dynamics since they reflect the accuracy of model construction in both systems. For example, the fishermen effort and harvest comparisons by season include both the social behavior of fishermen and the ecological behavior of salmon run-timing dynamics. The model uses the reconstructed salmon run-timing seasonal distributions to generate the salmon schools entering the simulation. The fishermen agents are generated using the recorded and the inferred counts. The interactions between the salmon and fishermen agents are measured in the model as seasonal harvest distributions using several measurements that include: cumulative harvest counts per stakeholder, catch per unit of effort of the stakeholders, and harvest by stakeholder groups relative to season run-size.

The model is validated by calculating the mutual correlations between the reported data by the management agencies (ADFG, Kenai Borough) and the measured distributions from the model's output. The sonar instrumented salmon counts are used to verify the cumulative impact of commercial set and drift gillnet fleet combined with the dipnet harvest. Figure 3a and b shows the validation of the individual stakeholder group behaviors and salmon behaviors in the context of dipnet effort and harvest by correlating measured data and the model's output within each system of the coupled systems dynamics. For each year, the model was seeded with the appropriate number of stakeholders and returning run dynamics to assess the model's accuracy at capturing observed social system dynamics and ecological dynamics. Figure 3a, b shows both intra-system dynamics metrics with correlation values $R^2 \geq 0.83$.

Coupled socio-ecological dynamics are tested by comparing the seasonal CPUE distributions of the stakeholder groups from the simulations with CPUE from the ADFG reports. Annual CPUE is a measurement of stakeholder effort and may be partially reflective of salmon abundance and other social and ecological factors affecting salmon harvest amounts (stakeholder or fleet efficiency, gear used, fishing location, run-timing dynamics, etc.). Figure 3 illustrates an example of social and ecological dynamics by comparing the recorded (c) ADFG dipnet CPUE with the recorded (d) model dipnet CPU for 2002–2014. The respective correlations of $R^2 = 0.77$ and $R^2 = 0.73$ for harvest and effort dynamics measure the model's ability to accurately capture the timing of salmon abundance and timing of dipnet effort, in addition to other factors that affect harvest mentioned above. In addition to assessing socio-ecological dynamics, CPUE validates the inference of the missing social system dynamics from the coupled-system dynamics. With data available to reconstruct run-timing dynamics, deviations of model CPUE from reported CPUE are due to run-timing error and/or social dynamics being our best heuristic effort. Large deviation from reported CPUE is then likely due to the inferred social dynamics. This allows for experimentation with different drivers to re-assert the social dynamics until model CPUE variation lies within the bounds of variability introduced by run-timing dynamics. Set gillnet dynamics in the model are also being constructed using human developed heuristics.

To test the model's ability to generalize the system behavior of both social and ecological systems, we executed the model independently with a new random seed each time generating the salmon run-timing dynamics by randomly sampling the biased roulette wheel of four run-timing prototypes of the past 35 years. Overall salmon abundance was randomly determined from the range of values typically observed since 2002 for purposes of comparing generalized run results with previous validation-based runs from 2002 to 2014 (2–6.5 million sockeye salmon returning, results not shown).

4 Behavioral Sensitivity Analysis

We conducted a series of stakeholder engagement meetings to solicit input from the resource managers, scientists, policy-makers, and economists to formalize plausible future scenarios. These scenarios are then translated to a range of possible values to

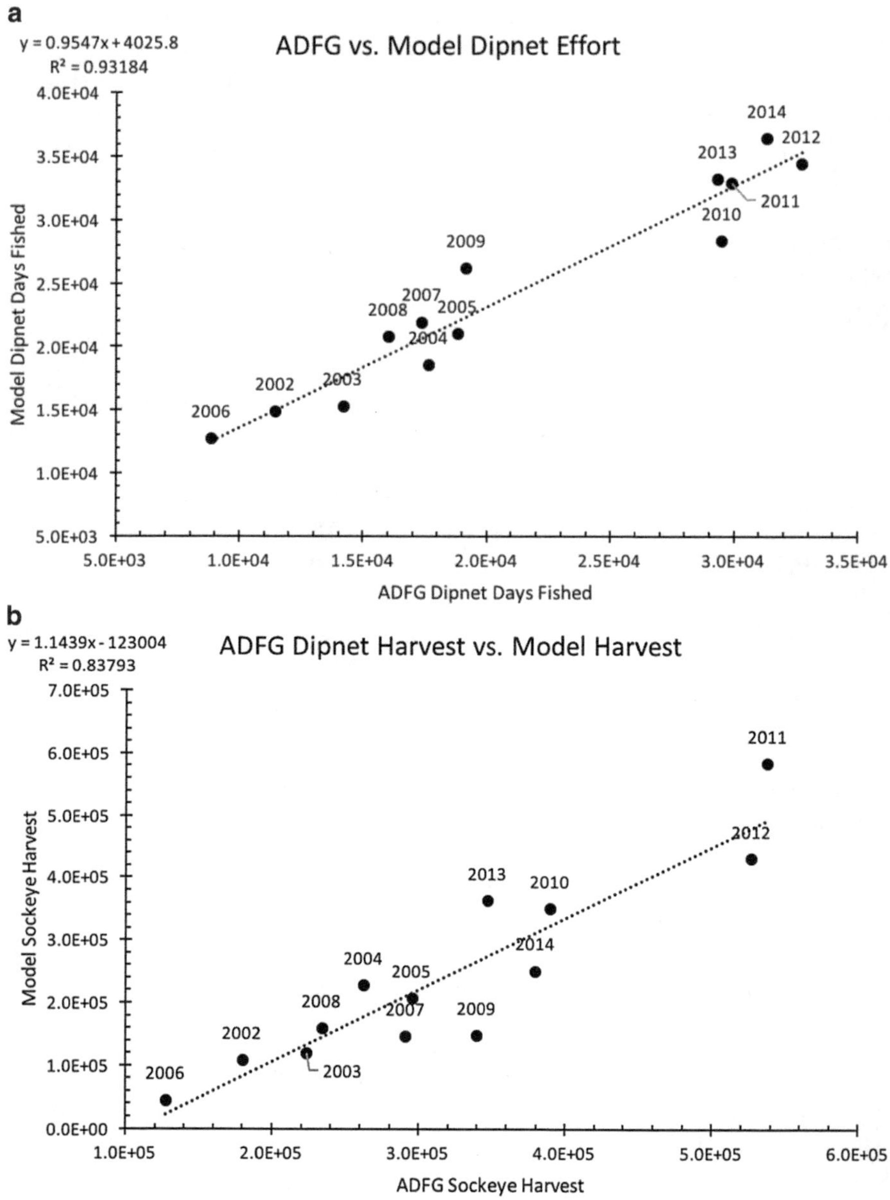

Fig. 3 Validating Social System Dynamics of (**a**) dipnet effort and (**b**) dipnet harvest. Seasonal effort and harvest data from the model output and historical records were cross-correlated to validate the social dynamics of the model. Ecological System Dynamics are also reflected in dipnet harvest (**b**). Sub-figures (**c**) and (**d**) show validation of the coupled system dynamics using (**c**) dipnet CPUE from data and (**d**) the model's output. The seasonal data of CPUE used permit days fished versus harvest

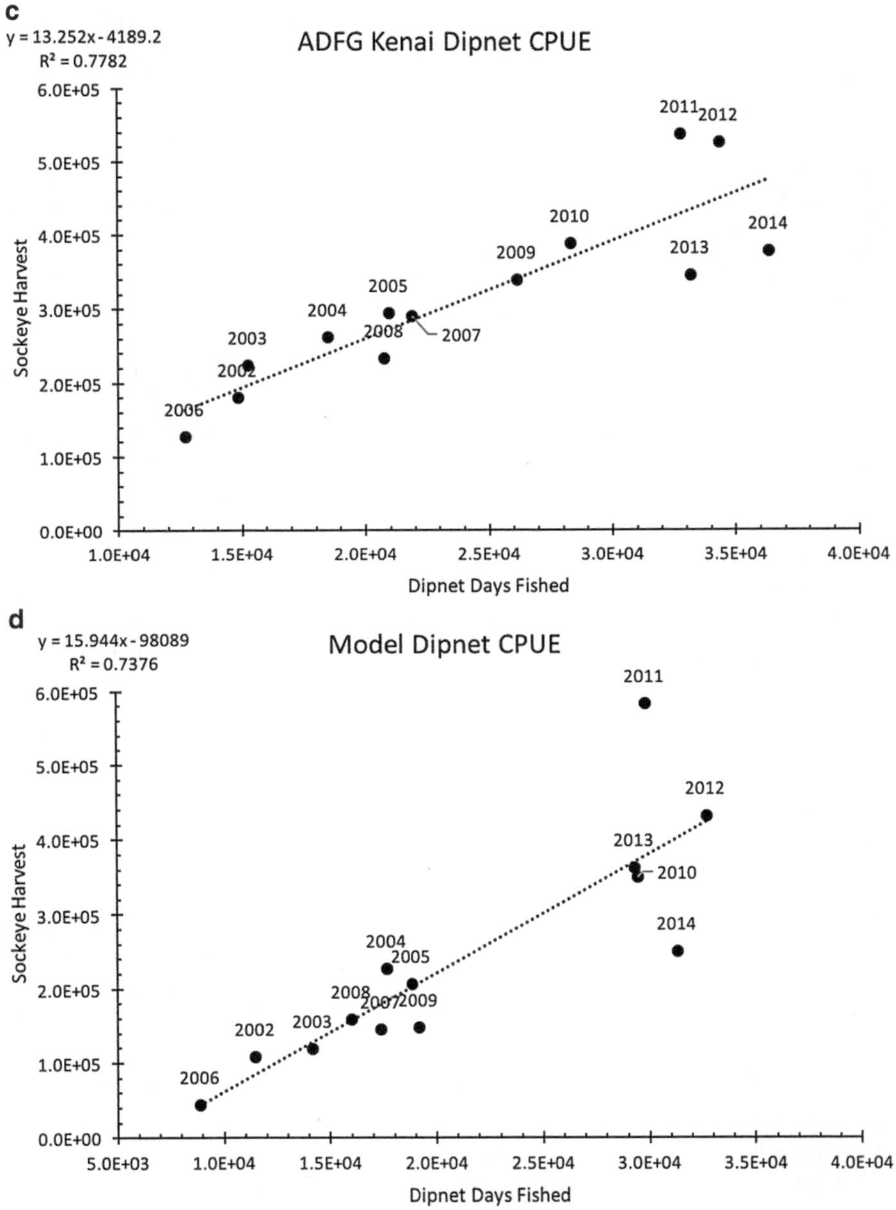

Fig. 3 (continued)

the model's variables. The model is then executed for all possible value permutations while the ABM's agent behaviors are recorded. The model executions are evaluated by statistical analysis of the model's executions and by comparing the behaviors of the model's agents.

Measuring model outcomes with simple metrics loses the information about how the goal was met, or how the nature of interactions between the model's agents changed to produce the system-wide (outcomes) dynamics. Visually inspecting each model behavior is infeasible for the combinatorial parameter space. We developed a statistical based toolbox called Geometry of Behavioral Spaces (GOBS) that records agents' behaviors independent of the knowledge of the parameter space that drives the model and produces a state-space transition network that characterizes the agent behaviors [7].

A model simulation area is tessellated into regular orthogonal or irregular Voronoi cells to compress space. Next, each agent logs its movement when crossing a cell boundary. The statistically based framework analyzes the recorded agent trajectories and detects common behaviour patterns (behavioral primitives) in the recorded trajectories. The final step is to construct a probabilistic state-space behavioral network that captures how likely an agent in a given behavioral primitive is to stay in the same pattern of behavior or to transition to a different behavioral primitive. Analyzing a model execution using the GOBS framework provides a tunable multi-scale view of agent behaviors, quantitative and qualitative comparison of multiple models, and a behaviour based (rather than a parameter or a simple metric) analysis. The GOBS computational framework is used to understand the adaptive capacity and conduct the sensitivity analysis of the model's coupled system dynamics. The framework analyzes how the agents behavior changed with the alteration of parameter values.

5 Conclusions and Future Work

We described a construction of a high fidelity ABM from data sources with high diversity, unknown accuracy, and various reporting frequencies. We constructed collections of temporal distributions that described the individual social and ecological system dynamics as well as the coupled system dynamics. The regressions between the data collections representing instrumented measurements and the model's outputs measure the accuracy of the model's construction in generalizing the socio-ecological systems. The data collections from instrumented measurements often contained multiple distributions describing the same observed phenomena. By cross-correlating these equivalency distributions, we performed manual ensemble learning to establish trustworthiness of each source distribution.

Data collection initiatives at Kenai Peninsula are crucial to understand the fisheries dynamics. The biophysical data collection sites generate data that is being reported on a daily to annual basis, and as new algorithms are developed the historical data are continuously recomputed and adjusted. Multiple data sources report the same phenomena, which allowed us to isolate the individual system dynamics, infer the non-existent system dynamics, validate the trustworthiness of data-sets, and define the socio- and bio-spatial boundary of the model study area.

The ABM construction generalized the longitudinal coupled social and ecological trends for further studies of system sensitivity or impact of individual drivers of change on the fisheries.

The ABM is an adaptable, data-driven platform that can be instantiated with data from another fishery for scenario based studies. The ecological changes observed at different locations will be used to alter existing dynamics to study systems' adaptive capacity and sensitivity. The utility of the ABM as a decision support tool will allow fishery managers to test plausible future scenarios by using observed changes to spatial-temporal dynamics from a different fishery to Kenai.

Future research includes implementing plausible future scenarios identified in a series of participatory stakeholder engagement meetings to understand how the coupled system dynamics will change in each scenario. The social scenarios include using dipnetters as a means for managing escapement, alteration of commercial gillnet fishing gear for reducing non-target species by-catch, and using sports fishermen as a means for controlling escapement. The ecological scenarios include compressing the salmon run duration by 2 weeks while maintaining the abundance and inversely keeping the overall dynamics while reducing the overall salmon abundance. Finally, we will use the statistical toolbox (Geometry of Behavioral Spaces Framework) to analyze agent behavior to understand system outcomes and model changes in terms of agent behaviors.

Acknowledgements This work was funded by Alaska EPSCoR NSF award #OIA-1208927. The authors would also like to thank all project collaborators for their comments and suggestions especially to Dr. Dan Rinella, Molly McCarthy, and ADFG Staff for sharing their incredible wealth of knowledge of Kenai Fisheries.

References

1. 2015 Kenai River Dipnet Fishery (2015). http://www.ci.kenai.ak.us/. Accessed 01 Apr 2016
2. Alaska Department of Fish and Game (2016). http://www.adfg.alaska.gov. Accessed 01 Apr 2016
3. Allen PM, McGlade JM (1986) Dynamics of discovery and exploitation: the case of the scotian shelf groundfish fisheries. Can J Fish Aquat Sci 43(6):1187–1200
4. Barclay AW, Habicht C, Tobias T, Willette TM, Templin WD, Hoyt HA, Chenoweth EL Genetic stock identification of Upper Cook Inlet sockeye salmon harvest, 2005–2008, 2009, 2010, 2011. Alaska Department of Fish and Game, Division of Sport Fish, Research and Technical Services, 2010, 2013, 2014
5. Branch TA, Hilborn R, Haynie AC, Fay G, Flynn L, Griffiths J, Marshall KN, Randall JK, Scheuerell JM, Ward EJ, Young M (2006) Fleet dynamics and fishermen behavior: lessons for fisheries managers. Can J Fish Aquat Sci 63(7):1647–1668
6. Cabral RB, Geronimo RC, Lim MT, Aliño PM (2010) Effect of variable fishing strategy on fisheries under changing effort and pressure: An agent-based model application. Ecol Model 221(2):362–369
7. Cenek M, Dahl SK (2016) Geometry of behavioral spaces: A computational approach to analysis and understanding of agent based models and agent behaviors. Chaos 22(11):113107

8. Dupuis A, Willette M, Barclay A (2011) Migratory timing and abundance estimates of sockeye salmon into Upper Cook Inlet, Alaska, 2010. Alaska Department of Fish and Game, Division of Sport Fish, Research and Technical Services
9. Effort and Catch per Unit Effort (2016). http://www.fao.org/. Accessed 01 Apr 2016
10. Quinn TP (2005) The behavior and ecology of Pacific salmon and trout. University of Washington Press, Seattle
11. Ricker WE (1958) Handbook of computations for biological statistics of fish populations. Fisheries Research Board of Canada, Ottawa
12. Shields P, Dupuis A (2015) Upper cook inlet commercial fisheries annual management report, 2011–2014. Alaska Department of Fish and Game, Division of Sport Fish and Commercial Fisheries, Fishery Management Reports No. 2014:15–20, 2013:13–49, 2012:13–21, 2011:12–25 Soldotna
13. Tobias T, Willette M, Tarbox K (2004) An estimate of total return of sockeye salmon to upper cook inlet, alaska 1976–1998, 1976–2003. Alaska Department of Fish and Game, Division of Commercial Fisheries, Regional Information Reports 1999:2A99–11, 2004:2A04-11, Anchorage
14. Van Putten IE, Kulmala S, Thébaud O, Dowling N, Hamon KG, Hutton T, Pascoe S (2012) Theories and behavioural drivers underlying fleet dynamics models. Fish Fish 13(2):216–235

Leveraging Coupled Agent-Based Models to Explore the Resilience of Tightly-Coupled Land Use Systems

Patrick Bitterman and David A. Bennett

Abstract This chapter argues that agent-based models (ABMs) possess an inherent advantage for modeling and exploring the general and specified resilience of social-ecological systems. Coupled systems are often complex adaptive systems, and the ability of ABMs to integrate heterogeneous actors, dynamic couplings, and processes across spatiotemporal scales is vital to understanding resilience in the context of complexity theory. To that end, we present the results of a preliminary stylized model designed to explore resilience concepts in an agricultural land use system. We then identify strengths and opportunities for further ABM development, and outline future work to integrate empirically-parameterized agent behavioral rules with robust biophysical models to explore resilience and complexity.

Keywords Resilience • Agent-based modeling • Complexity • Adaptive capacity

1 Introduction

Change in social-ecological systems is inevitable. These changes may be a result of new or increasing environmental pressures (e.g., increased precipitation variability), or a response to technological, political, or economic developments. Whether gradual or abrupt, purposeful or unintended, shifts among alternative system states can lead to new configurations of social and physical landscapes, accompanied by changes in key properties (e.g., stability, sustainability, environmental quality, productivity) that describe a system's function, its relationships with other systems, and its desirability to humans. The social-ecological resilience paradigm provides a useful framework to conceptualize the breadth of potential system states, transitions among them, and their impact on environmental outcomes and human well-being.

P. Bitterman (✉)
University of Vermont, Burlington, VT, 05405, USA
e-mail: patrick.bitterman@uvm.edu

D.A. Bennett
University of Iowa, Iowa City, IA, 52242, USA

L. Perez et al. (eds.), *Agent-Based Models and Complexity Science in the Age of Geospatial Big Data*, Advances in Geographic Information Science,
DOI 10.1007/978-3-319-65993-0_2

However, in complex, adaptive, and tightly-coupled systems with strong linkages between environmental function and social well-being, this view of resilience has proven difficult to operationalize due to the complicated and uncertain nature of human decision-making. Agent-based modeling techniques, and more broadly coupled modeling and GIScience, are well-positioned for operationalizing this complex view of resilience due to their ability to handle heterogeneous actors, adaptation, dynamic couplings, and processes that reach across time and space and spatiotemporal scales. CHANS-oriented agent-based models (ABMs) have been shown useeful in modeling non-linear processes, feedbacks, lags, heterogenity, and general resilience in complex, coupled systems [1]. The objective of this chapter is to demonstrate how agent-based models, when coupled with social-ecological resilience theory, can efficiently and effectively explore alternative system states and guide policymaking. In this chapter, we describe initial work designed to explore some central concepts from resilience theory using a simplified ABM of agricultural land use in shifting policy and climatic contexts. We then discuss hurdles to fully implementing such models in the search for resilient and sustainable states, and identify opportunities for future work to address those obstacles.

2 Resilience and Complexity

Systems in which human activity and ecological function are tightly linked, whether termed social ecological systems (SES) [2], coupled human and natural systems (CHANS) [3], or otherwise [4], exhibit complexity resulting from the linkages and feedback processes among systems. Traditional models of resilience in ecosystems have typically employed differential equations and isoclines to determine alternative stable states created by environmental and anthropogenic perturbations [5, 6]. Geographic applications generally focus on environmental hazards [7], metrics of exposure and response [8, 9], and adaptive management [10, 11]. However, most of the current geographic methods provide only snapshots in time of a system defined, for example, by US Census variables and enumeration units, and do not incorporate underlying key processes. Although these different epistemological perspectives and disciplinary traditions vary in their approaches to addressing overlapping questions of resilience, sustainability, and vulnerability, there remains general agreement that their central challenges lie in the analysis of the linkages among systems [2].

Resilience has taken on many meanings depending on application or field, and has been capably and thoroughly reviewed in many contexts [8, 12–14]. More recent work has acknowledged a multiplicity of resilience definitions, and effectively argued for a more open and holistic, though structured, paradigm for the purposes of resilience assessment and measurement [15, 16]. For example, resilience thinking can be applied in qualitative assessments of high-level policy and governance in rural social, economic, and environmental changes [17]. More narrowly, ecological resilience generally addresses the ability of a system to absorb a disturbance and maintain its structure, function, and identity [18], meaning that if a new state is reached and maintained, the resilience of the previous system state was exceeded. While termed "ecological" due to disciplinary roots, this resilience paradigm integrates the bidirectional feedbacks linking human wellbeing and environmental function. There are essentially two broad methodological approaches to measuring this form of resilience: (1) the amount of disturbance to which a system can be subjected, and (2) the length of post-perturbation recovery time before dynamic equilibrium is restored [13]. To operationalize resilience in a broader, integrative social-ecological perspective, we must place the system's current and potential states, as well as relevant perturbations, in a specified context. Such a view of resilience sets SES within a complex hierarchy of nested, connected, adaptive, and constantly changing systems, termed *panarchy* [19]. In this panarchy, connected systems pass materials, energy, or information across scales and at different stages of the adaptive process, influencing their resilience. These are aspects that a specified view of resilience focused on "of what to what" questions attempts to address [20]. In contrast to a specified approach, the general view of resilience considers the unknown disturbance, and emphasizes the complex interactions, uncertain outcomes, and the potential for surprise [21]. Connections between general and specified views are found in the adaptive capacity of SES, also termed as adaptability [22, 23]. Adaptability refers to the ability of actors (e.g., individuals, groups, institutions) within the system to manage resilience and shape outcomes at various scales within the nested systems. It should be emphasized, however, that resilience is not necessarily a desirable property, as efforts to increase it may inadvertently lead to "lock in traps" and maladaptive states [24, 25]. Further, whether resilience is considered desirable is an inherently subjective judgement, and should consider the many complex ways human well-being is connected to ecological function and socio-political structures and dynamics.

A social-ecological perspective of resilience acknowledges that SES are complex adaptive systems (CAS) [26, 27]. CAS possess an evolving structure, aggregate behavior, and actors with the ability to anticipate consequences of their actions [28]. The resilience of SES may stem from the aggregate behavior of individuals, the effects of top-down policy instruments, or the transfer of materials or capital from one location or scale to another. Therefore, resilience is inherently a geographic problem and complexity problem, as the spatial structure of both the system and the disturbance can greatly influence the impact, response, and recovery to a perturbation [29]. Similarly, the network topology of connected system components at finer scales, and of whole systems at broader scales, can affect the flow of coping capacity and recovery functions, directly resulting in shifts in resilience [30]. Along these connections collective behavior, information processing, and adaptation strategies form, generating non-linear processes from the resultant legacy effects, path dependency, and spatial and temporal lags in the SES [31–34]. Accordingly, we argue that ABMs are particularly appropriate for exploring the resilience of coupled systems. ABMs can represent heterogeneous actors on a landscape and at different scales [35], can integrate context into decision-making [36], and through repeated simulations, can produce metrics identifying expected outcomes and unlikely shifts in state [37, 38]. More importantly, ABMs can facilitate simultaneous examination of the specified resilience of individuals and groups to spatially-explicit perturbations, and of general resilience emerging from agent-agent and agent-environment interactions [39, 40]. What follows is a description of a stylized model designed to explore the resilience and outcome spaces of a small watershed in Eastern Iowa, with the intent to expand the model in spatial scale and complexity based on lessons learned from the exploratory process and survey data.

3 Preliminary Modeling Methods

Our area of interest is the Iowa-Cedar River Basin (ICRB), which covers approximately 33,000 km^2 of Eastern Iowa and a small portion of southern Minnesota in the Midwest U.S. Land use in the basin is dominated by intensive agriculture, with 70% of land use devoted to either corn or soybean production (Fig. 1). The ICRB has recently experienced multiple extreme flood and drought events [41], and is at the center of regional and national issues concerning the economic and environmental costs of nutrient run-off, soil loss, and water pollution. To limit computational overhead, our preliminary study focuses on the Clear Creek Watershed (CCW), a highly-studied and heavily instrumented [42–44] sub-watershed in the ICRB.

To explore resilience within the SES, we constructed a stylized ABM of farmer land use, economic and environmental drivers, and outcomes coupled with relatively simple biophysical models of crop growth, run-off [45] and soil loss [46]. The model is custom software written in the Java programming language (JDK 1.8.0), and all analysis and visualization is performed in the R 3.2.3 software [47]. The

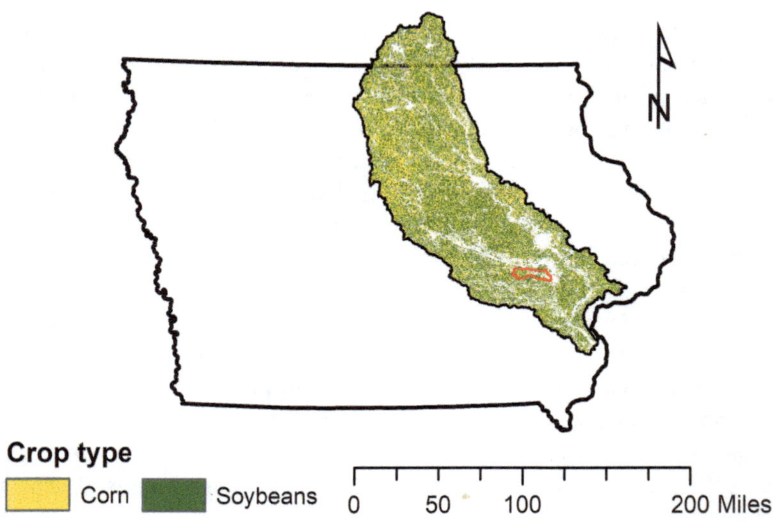

Fig. 1 Location of the Iowa-Cedar River Basin within the state of Iowa. Land use in the ICRB is dominated by corn and soybean agriculture, covering approximately 70% of all land in the basin. Location of the Clear Creek Watershed shown in the southern portion of the ICRB

purpose of the model was to better understand how broad-scale shifts in climate and policy might affect the performance and resilience of individual farmers, potentially affecting environmental outcomes and resilience at the watershed or basin scales. While a complete model description, documentation, and OpenABM.org link can be found elsewhere [48], we present here a brief sketch of the model's structure and function. Farm fields are the basic spatial unit of analysis in the model, and multiple fields comprise a farm, which is managed by a farmer. Each field is instantiated from common land unit data provided by the National Agricultural Survey Service (NASS). The primary class of agents in the model represents these farmers, each of whom is assigned a set of fields to manage using prescribed decision rules. In the absence of disturbance, the model proceeds in a straightforward manner. At an annual time step, farmer agents attempt to make profit-maximizing land use decisions by integrating information from past performance, market prices, and field characteristics. Crop prices are based on information from April 2015 [49], while switchgrass prices are instantiated at zero but are modified according to scenarios described later. Transition costs are estimated from Iowa State University extension data [50], and crop suitability ratings are used to provide yield estimates (USDA 2013). Farmer agents then implement their decisions, modifying field-level land use, and incurring costs. Fields then differentially accrue biomass according to land use and land suitability. Finally, farmer agents harvest biomass at the end of the season, sell at market prices, and balance their accounting ledgers, amassing profits (or losses) for their operation and collecting insurance indemnities on a portion of crop losses. If the profitability of a farmer agent is negative for three consecutive years, they are removed from the simulation. The model includes minor stochastic variability for crop yield [51], prices, and costs to approximate system uncertainty and variability.

The model, while parameterized with empirical land use data and realistic parameters, is not intended to be prescriptive, but rather to explore connections between the specified resilience of individual agents and general resilience concepts (e.g., diversity, coping capacity) at the system scale. The execution of the model as briefly described above and more fully in [48] provides basic estimates of system stability and sensitivity to initial conditions, as shown in Fig. 2. Here, we performed a grid search through an input space of commodity prices and subsidies to plot the distribution of the model's outcomes in a state space defined by mean net profitability per agent and landscape diversity, as measured by the Modified Simpson Diversity Index (MSIDI) [52]. MSIDI is one measure of landscape diversity (e.g., Shannon Diversity [53]), but we use MSIDI as its interpretation is more straightforward. The value of the MSIDI is defined as the negative natural logarithm of the probability two randomly selected patches belong to the same land cover type. Each point in the state space corresponds to a simulation outcome, and thus to one realization of the CCW landscape. The Fig. 2 state space is not a measure of resilience however, as the model at this point lacks any perturbations or functional coping mechanisms. The model is expanded by introducing scenarios that modify policy, economics, and climate by creating: (1) two levels of simulated drought perturbations, (2) the presence/absence of an artificial market for the

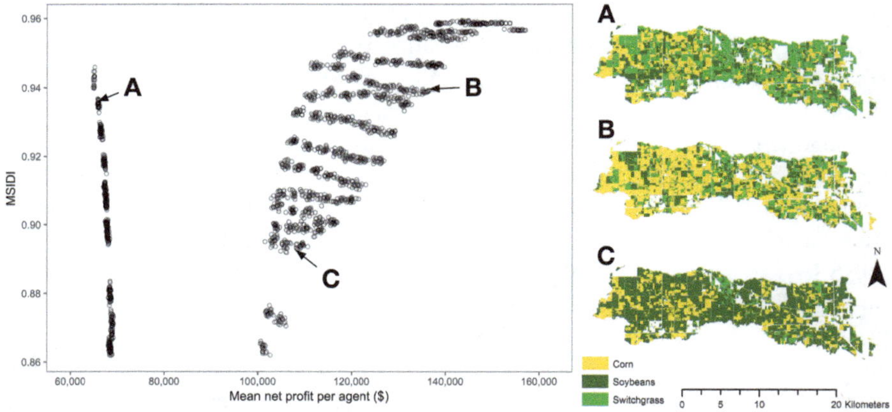

Fig. 2 A stability landscape of simulation outcomes and corresponding land use in the CCW. Two attractors are found in the state space. Higher MSIDI values indicate a more diverse landscape

cellulosic biofuel switchgrass, and (3) three levels of income reimbursement for losses via the federal crop insurance program. In combination, these modifications to the model produce 12 distinct scenarios, and simulations were performed 1000 times for each scenario, resulting in 12,000 total model runs. Once a model run reaches dynamic equilibrium, it is perturbed with a simulated drought, reducing the amount of biomass in each farm field. The spatiotemporal distribution of drought intensity is stochastically varied between runs of the model to produce a distribution of possible outcomes. We present the simulation results of one such scenario in the state space in Fig. 3, which is parameterized using the lowest level of insurance reimbursement, the presence of an artificial land use market for switchgrass, and relatively severe droughts. These simulation results provide preliminary insight into possible alternative states, and to the resilience of both farmer agents and landscape configuration within a given policy, economic, and climatic context.

4 Preliminary Results and Discussion

The state space of simulation outcomes in Fig. 2 shows the results of the 12,000 model runs along two dimension described in the previous section. Clusters of outcomes correspond to attractors within a stability landscape (i.e., regions in state space where the system is more likely to remain) [18]. Absent perturbation, and within the bounds of model design and our grid search through input space, these are the range of likely model outcomes. Two primary attractors are found, largely differentiated by the amount of corn and switchgrass on the landscape, a function of relative commodity prices and indicative of market thresholds.

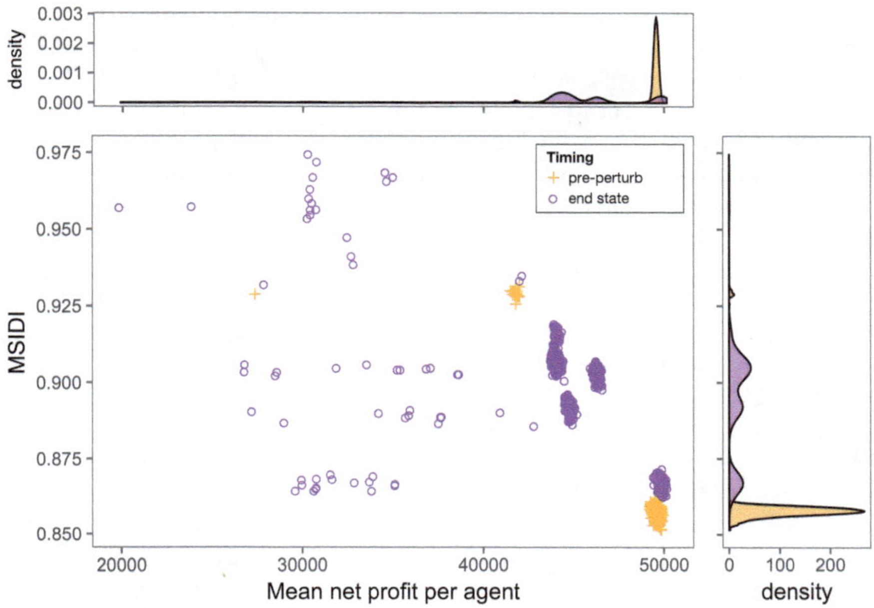

Fig. 3 State space locations for 1000 model runs of one scenario (low coping capacity, artificial switchgrass market, more severe droughts), classified by the pre-perturbation states (*orange crosses*) and end states (*purple circles*). The model generally occupies one area in state space prior to the application of the artificial droughts, indicative of a local dynamic equilibrium. Post-perturbation, system state is much more variable, indicating a loss of profitability and a more heterogeneous landscape

Within-group clustering is an artifact of both the steps through parameter space and the discretization of the landscape into parcels. If a finer grained search of input space were conducted, or if the size and shape of management units (i.e., fields) were more flexible, we would expect the state space within attractors to be nearly continuous.

In this space, each point corresponds to a realization of the actual landscape at the completion of a simulation. For example, one simulation with a set of initial conditions favoring switchgrass production results in the landscape in Fig. 2A, and is found in the left attractor in the state space. Conversely, alternative economic and climate configurations may be realized as landscape configurations in Fig. 2B and C, found in the right attractor. Though the landscapes in 2A and 2B have similar diversity (MSIDI) values, the configuration in 2B is much more profitable on a per-agent basis due to relative differences in crop costs and prices. In the case of 2C, market prices for corn and soybeans were substantially lower, shifting the market in favor of soybean production, reducing landscape diversity and profitability. The morphology of the state space is dependent on variable selection. While farmer profitability is desirable and more diverse landscapes are generally more resilient to climate variability due to increased functional redundancy [54], Fig. 2 excludes dimensions directly related to run-off, soil loss, or general ecosystem health and does not comprehensively capture the state of the system. While we set out to explore the specified resilience "of what, to what" questions in the context of agricultural land use and selected variables accordingly, we must also acknowledge the many unrepresented dimensions in the state space (e.g., environmental conditions). What appears to be an attractor in two-dimensions may in fact be multiple distinct basins in a higher dimensional space. Further, as SES move through time and are repeatedly perturbed, we must also consider the effects of path dependency in limiting the movement of SES among attractors, and its effects on the multifinality or equifinality of eventual end state(s). For example, two farmers affected by different types of perturbations (e.g., flood vs. drought) may make different adaptive decisions, thereby differentially affecting the economic costs of subsequent adaptations, resulting in a bifurcation point between these farmers. Conversely, we can imagine a case where changes in economic or environmental conditions result in a later convergence of these same farmers. The ability of ABMs to model SES structure (e.g., spatial, hierarchical, economic) in specific spatiotemporal contexts is a clear methodological advantage for the study of resilience.

The state space in Fig. 3 examines one of the scenarios described above to illustrate how the model reacts when subjected to artificial drought perturbations. In the figure, the orange crosses indicate the model locations at dynamic equilibrium and at the time step immediately preceding the perturbation. For this scenario parametrization, these locations are generally found in a single attractor in the bottom-right portion of the state space, though a second, smaller attractor is present as well. Once the model is perturbed, farmer agents react by consuming coping capacity in the form of crop insurance when losses exceed a given threshold. Some farmers go out of business, while those remaining update their profit expectations, thereby affecting their land use decisions in subsequent time steps, and moving the SES towards a new dynamic equilibrium (purple circles in Fig. 3). The heterogeneity in farmer response is seen in the distribution of post-perturbation end states in Fig. 3, which are more spread across the state space than pre-perturbation states. The new dynamic equilibria reached indicate alternative potential states, within the particular policy, economic, and disturbance context of this scenario.

While these state spaces identify potential states, the pre-and-post-perturbation outcomes are path dependent, and the state space is likely discontinuous between attractors. Further, "windows of opportunity" may open (and close), creating pathways to alternative states available only at particular moments or places. Finally, and in addition to numerical differences, these alternative states should be qualitatively evaluated with respect to stakeholder values and preferences. Despite its limitations, state space morphology can provide an initial guide to understanding under which scenarios the system may be more resilient. A more heterogeneous post-perturbation state space, for example, indicates a larger variance in model response to perturbation, and potentially lower resilience of the pre-perturbation attractors.

5 Future Work and the Promise of ABMs for Resilience

Our preliminary model was designed to explore the feasibility of utilizing a coupled ABM to understand the resilience and adaptive capacity in the ICRB agricultural SES. The model was successful in plotting individual and aggregate farmer response to perturbations in particular policy, economic, and climatic contexts, and generated outcome state spaces analogous to theoretical stability landscapes. However, by design the model did not leverage the full strength of intelligent agents, which would include changing adaptive strategies, agent communication, and cooperative behavior. Priorities for quantifying and operationalizing resilience in a spatial context have recently been proposed [55], and in this section we identify strengths and opportunities for ABMs, and geographic modeling techniques more broadly, to address questions of specified and spatial resilience.

The management of resilience necessarily requires adaptive capacity, which can be manifested at individual, group, or system scales. For example, individual farmer agents might adapt their land use, practices, or goals within particular constraints. Similarly, government agencies might adapt their policies and offer (dis)incentives for particular system level outcomes (e.g., water quality standards). Our stylized model demonstrated that when these decisions or policies are relatively static, the determination of alternative states is straightforward, if subject to simplifying assumptions. However, to more fully understand SES resilience, models must incorporate those key linkages and feedbacks within and among social and ecological systems. Those same institutional policies, for example, might generate collective behavior in farmers to self-organize to meet regulatory requirements, or they might fracture a community and promote further competition for common pool resources. However, the adaptive capacity of actors or groups is constrained in various ways that limit agents' ability to adapt. For example, a knowledge of system function and its potential for future change is required for purposeful adaptation, and the ability of actors to learn is necessary to increasing system-scale adaptive capacity [56, 57]. Access to resources and capital (social and natural) is required to plan, implement, and manage an adaptation. This access is often unequally distributed, and is shaped by the social and institutional context [10] in which agents are placed. While our stylized model did not include direct agent interactions, these dynamics can be at least partially captured by modeling social networks among agents or agent typologies, which can affect land management strategies and model outcomes [58]. Finally, adaptive capacity is limited by individual and group willingness to implement and accept an adaption. These impediments might be based on values and beliefs [59] or cognitive biases and risk perception [60], among others. ABMs for resilience must incorporate these constraints and provide mechanisms within the model for adaptive strategies to change if the constraints are altered.

There exist, of course, many examples of ABMs for land use modeling and environmental management [61–63]. There are methods for agents to learn from their experiences, other agents, or their environment [64], and the abstraction of decision algorithms within an object-oriented framework simplifies the modeling process. We simply argue that ABMs, GIScience, and geography more broadly, possess an inherent advantage to operationalizing resilience in a way that integrates spatially-explicit human and environmental dynamics in a multi-scale context and across space and time. Modeling allows for experimentation, and excepting natural experiments, provides the sole method for exploring potential effects of, for example, new climatic regimes or policy instruments. Further, although the ecological resilience literature has identified methods for identifying thresholds and critical transitions in ecosystems [65, 66], coupled ABMs can help assess linked social and environmental outcomes, and perhaps determine bifurcation points that leads to alternative system states [58].

Future work will build on the stylized model presented above, and integrate data on farmer constrains on adaptation to perturbations. From a mailed survey of 1200 farmers, we collected data on farmer risk perception, experiences, and potential adaptions for each of three potential perturbations (excess rain, drought, and agricultural policy). We also collected data on seven constraints for each potential adaptation. From these data and demographics (e.g., farm size, income, climate change views), we have generated farmer profiles and corresponding distributions of likely responses to disturbances, and have identified potential constraints on farmer decisions. For example, Fig. 4 shows the results of a survey question that asked "After implementing practices to improve water quality . . . how many years would you expect to wait before water quality improvements are noticeable?" The majority of farmers expect improvements in water quality to occur in fewer than 4 years, which is far shorter than models have estimated in similar watersheds [67]. This and similar misconceptions about ecological dynamics can limit willingness to adapt, or

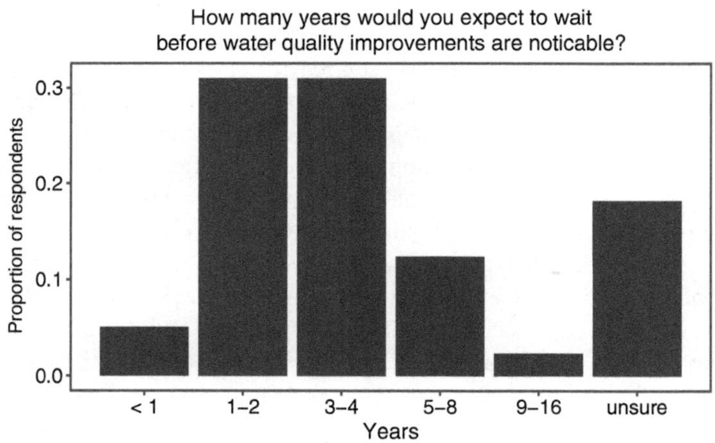

Fig. 4 The distribution of farmer responses (N = 258) to the question: "After implementing practices to improve water quality (for example, reducing nitrogen application rates, installing filter strips), how many years would you expect to wait before water quality improvements are noticeable?" The majority (67%) of respondents expect to see water quality improvements in time scales that are unlikely given the nitrogen legacy on the landscape

create lock-in traps where farmers revert to past management practices when they fail to see environmental benefits on expected or acceptable timescales.

Work is ongoing to extend our model to integrate these survey data with more robust climate and biophysical models to more tightly model the feedback mechanisms that drive land use and adaptation. Using these survey data, we will empirically parameterize agent behavior rules that integrate: (1) a suite of climate models reaching to the year 2070, (2) shifts in crop insurance policy introduced in the 2014 US Farm Bill, and (3) recent price fluctuations affecting farmer profitability and use of marginal lands. Further, through novel couplings between the ABM and the Soil Water and Assessment Tool (SWAT), we will generate artificial, spatially-explicit perturbations to search for "levers" that force the model across thresholds and to new states. Through experimentation on the virtual landscape, we will utilize the specified resilience framework to plot system responses over time and over many repeated simulations, identify the relevant structures and processes that create bifurcation points in system trajectory.

6 Conclusion

Resilience emerges from interactions that span the panarchy of components, both connected to, and nested within, a complex adaptive system [19]. A specified view of resilience requires the consideration of the spatial structure of both system panarchy and perturbation, necessitating a spatially-explicit approach rooted in Geography. However, an ability to model transformational change in adaptive strategies and to generate novel couplings within and among social and ecological systems, remains elusive yet necessary to understanding SES resilience. Our stylized model of agricultural land use in the CCW demonstrates how a relatively simple ABM can couple social and environmental models to plot the range of likely system responses to perturbation. Though the model situates farmer land use decisions in the context of a given scenario, it does not fully leverage the strengths of multi-scale agent-based model. For example, if agents are to change adaptive strategies, rather than simply modify equation coefficients, then the constraints on adaptation must be considered in model design, and the feedback mechanisms that modify those constraints must be explicit. Further, the spatial heterogeneity of agent capacity, and spatial structure of sources of coping capacity must be included. Stakeholder engagement can incorporate local knowledge not only to better understand system processes and produce improved models, but can also help identify those SES states desirable to system actors [68]. While models of specified resilience should recognize the diversity of agents, opinions, and motivations, and address power dynamics underpinning the "resilience of what, to what, and who decides" questions [69], preferred states may exist near thresholds, require significant inputs to maintain, or reduce sustainability in unexpected ways. ABMs can be an important tool to weigh the resilience, performance, and equity of potential outcomes for all parties, domains, and scales.

Acknowledgements This work is funded in part by NSF Coupled Natural Human Systems Award #1114978, People, Water, and Climate: Adaptation and Resilience in Agricultural Watersheds.

References

1. An L, Zvoleff A, Liu J, Axinn W (2014) Agent-based modeling in coupled human and natural systems (CHANS): lessons from a comparative analysis. Ann Assoc Am Geogr 104: 723–745
2. Ostrom E (2009) A general framework for analyzing sustainability of social-ecological systems. Science 325:419–422
3. Liu J, Dietz T, Carpenter SR, Folke C, Alberti M, Redman CL, Schneider SH, Ostrom E, Pell AN, Lubchenco J, Taylor WW, Ouyang Z, Deadman P, Kratz T, Folke C, Provencher W (2007) Coupled human and natural systems. Ambio 36:639–649
4. Binder CR, Hinkel J, Bots PWG, Pahl-Wostl C (2013) Comparison of frameworks for analyzing social-ecological systems. Ecol Soc 18:26
5. Grimm V, Schmidt E, Wissel C (1992) On the application of stability concepts in ecology. Ecol Model 63:143–161
6. Scheffer M, Carpenter S, Foley JA, Folke C, Walker B (2001) Catastrophic shifts in ecosystems. Nature 413:591–596
7. Cutter SL, Finch C (2008) Temporal and spatial changes in social vulnerability to natural hazards. Proc National Acad Sci USA 105:2301–2306
8. Cutter SL, Burton CG, Emrich CT (2010) Disaster resilience indicators for benchmarking baseline conditions. J Homeland Secur Emerg Manag 7:1–24
9. Lam NS-N, Qiang Y, Arenas H, Brito P, Liu K-B (2015) Mapping and assessing coastal resilience in the Caribbean region. Cartogr Geogr Inf Sci 42:315–322
10. Adger WN, Vincent K (2005) Uncertainty in adaptive capacity. Compt Rendus Geosci 337:399–410
11. Anderies JM, Folke C, Walker B, Ostrom E (2013) Aligning key concepts for global change policy: robustness, resilience, and sustainability. Ecol Soc 18:8
12. Zhou H, Wang J, Wan J, Jia H (2010) Resilience to natural hazards: a geographic perspective. Nat Hazards 53:21–41
13. Morecroft MD, Crick HQP, Duffield SJ, Macgregor NA (2012) Resilience to climate change: translating principles into practice. J Appl Ecol 49:547–551
14. Baggio JA, Baggio JA, Brown K, Brown K, Hellebrandt D, Hellebrandt D (2015) Boundary object or bridging concept? A citation network analysis of resilience. Ecol Soc 20:2
15. Davidson JL, Jacobson C, Lyth A, Dedekorkut-Howes A, Baldwin CL, Ellison JC, Holbrook NJ, Howes MJ, Serrao-Neumann S, Singh-Peterson L, Smith TF (2016) Interrogating resilience: toward a typology to improve its operationalization. Ecol Soc 21:27
16. Quinlan AE, Berbés-Blázquez M, Haider LJ, Peterson GD (2015) Measuring and assessing resilience: broadening understanding through multiple disciplinary perspectives. J Appl Ecol 53:677–687
17. Schouten MAH, van der Heide CM, Heijman WJM, Opdam PFM (2012) A resilience-based policy evaluation framework: application to European rural development policies. Ecol Econ 81:165–175
18. Walker B, Carpenter S, Holling CS, Kinzig A (2004) Resilience, adaptability and transformability in social–ecological systems. Ecol Soc 9:5
19. Holling CS (2001) Understanding the complexity of economic, ecological, and social systems. Ecosystems 4:390–405
20. Carpenter S, Walker B, Anderies JM, Abel N (2001) From metaphor to measurement: resilience of what to what? Ecosystems 4:765–781
21. Folke C (2016) Resilience (republished). Ecol Soc 21:44

22. Engle NL (2011) Adaptive capacity and its assessment. Glob Environ Chang 21:647–656
23. Walker B, Salt D (2006) Resilience thinking: sustaining ecosystems and people in a changing world. Island Press, Washington, D.C
24. Allison HE, Hobbs RJ (2004) Resilience, adaptive capacity, and the lock-in trap of the western Australian agricultural region. Ecol Soc 9:3
25. Carpenter SR, Brock WA (2008) Adaptive capacity and traps. Ecol Soc 13:40
26. Gunderson LH, Carpenter S, Folke C, Olsson P, Peterson G (2006) Water RATs (resilience, adaptability, and transformability) in lake and wetland social-ecological systems. Ecol Soc 11:16
27. Folke C (2006) Resilience: the emergence of a perspective for social–ecological systems analyses. Glob Environ Chang 16:253–267
28. Holland JH (1992) Complex adaptive systems. Daedalus 121:17–30
29. Cumming GS (2011) Spatial resilience in social-ecological systems. Springer, London
30. Janssen MA, Bodin Ö, Anderies JM, Elmqvist T, Ernstson H, Mcallister RRJ, Olsson P, Ryan P (2006) Toward a network perspective of the study of resilience in social-ecological systems. Ecol Soc 11:15
31. Mitchell M (2009) Complexity a guided tour. Oxford University Press, New York
32. Malanson GP (1999) Considering complexity. Ann Assoc Am Geogr 89:746–753
33. Liu J, Dietz T, Carpenter SR, Alberti M, Folke C, Moran E, Pell AN, Deadman P, Kratz T, Lubchenco J, Ostrom E, Ouyang Z, Provencher W, Redman CL, Schneider SH, Taylor WW (2007) Complexity of coupled human and natural systems. Science 317:1513–1516
34. Bennett D, McGinnis D (2008) Coupled and complex: human–environment interaction in the greater yellowstone ecosystem, USA. Geoforum 39:833–845
35. Parker DC, Manson SM, Janssen MA, Deadman P, Hoffmann MJ (2003) Multi-agent systems for the simulation of land-use and land-cover change: a review. Ann Assoc Am Geogr 93:314–337
36. Tang W, Bennett DA (2010) The explicit representation of context in agent-based models of complex adaptive spatial systems. Ann Assoc Am Geogr 100:1128–1155
37. Brown DG, Page S, Riolo R, Zellner M, Rand W (2005) Path dependence and the validation of agent-based spatial models of land use. Int J Geogr Inf Sci 19:37–41
38. Parker DC, Hessl A, Davis SC (2008) Complexity, land-use modeling, and the human dimension: fundamental challenges for mapping unknown outcome spaces. Geoforum 39:789–804
39. Guzy MR, Smith CL, Bolte JP, Hulse DW, Gregory SV (2008) Policy research using agent-based modeling to assess future impacts of urban expansion into farmlands and forests. Ecol Soc 13:37
40. Schouten M, Opdam P, Polman N, Westerhof E (2013) Resilience-based governance in rural landscapes: experiments with agri-environment schemes using a spatially explicit agent-based model. Land Use Policy 30:934–943
41. Mallya G, Zhao L, Song XC, Niyogi D, Govindaraju RS (2013) 2012 midwest drought in the United States. J Hydrol Eng 18:737–745
42. Schilling K, Streeter M, Hutchinson K, Wilson C, Abban B, Wacha K, Papanicolaou A (2015) Effects of land cover on streamflow variability in a small iowa watershed: assessing future vulnerabilities. Am J Environ Sci 11:186–198
43. Papanicolaou AN, Wacha KM, Abban BK, Wilson CG, Hatfield JL, Stanier CO, Filley TR (2015) From soilscapes to landscapes: a landscape-oriented approach to simulate soil organic carbon dynamics in intensively managed landscapes. J Geophys Res Biogeosci 120(11):2375–2401
44. Ding D, Bennett D, Secchi S (2015) Investigating impacts of alternative crop market scenarios on land use change with an agent-based model. Landscape 4:1110–1137
45. Budyko MI (1958) The heat balance of the Earth's surface. Office of Technical Service, U.S. Department of Commerce, Washington D. C
46. Renard KG, Foster GR, Weesies GA, Porter JP (1991) RUSLE: revised universal soil loss equation. J Soil Water Conserv 46(1):30–33

47. R Core Team: R, https://R-project.org
48. Bitterman P, Bennett DA (2016) Constructing stability landscapes to identify alternative states in coupled social-ecological agent-based models. Ecol Soc 21:21
49. AGWEB: Cash Grain Bids, https://www.agweb.com/markets/cash-grain-bids
50. Iowa State University (2016) Extension and outreach: estimated costs of crop production in Iowa–2016
51. Bakhsh A, Jaynes DB, Colvin TS, Kanwar RS (2000) Spatio-temporal analysis of yield variability for a corn-soybean field in Iowa. Trans ASAE 43:31–38
52. Pielou EC (1975) Ecological diversity. Wiley-Interscience, New York
53. Shannon CE, Weaver W (1949) The mathematical theory of communication. University of Illinois Press, Urbana
54. Elmqvist T, Folke C, Nystrom M, Peterson G, Bengtsson J, Walker B, Norberg J (2003) Response diversity, ecosystem change, and resilience. Front Ecol Environ 1:488–494
55. Allen CR, Angeler DG, Cumming GS, Folke C, Twidwell D, Uden DR (2016) Quantifying spatial resilience. J Appl Ecol 53:625–635
56. Brooks N, Neil Adger W, Mick Kelly P, Kelly PM (2005) The determinants of vulnerability and adaptive capacity at the national level and the implications for adaptation. Glob Environ Chang 15:151–163
57. Nelson DR, Adger WN, Brown K (2007) Adaptation to environmental change: contributions of a resilience framework. Annu Rev. Environ Resour 32:395–419
58. Bennett DA, Tang W, Wang S (2011) Toward an understanding of provenance in complex land use dynamics. J Land Use Sci 6:211–230
59. Moser SC, Ekstrom JA (2010) A framework to diagnose barriers to climate change adaptation. PNAS 107:22026–22031
60. Grothmann T, Patt A (2005) Adaptive capacity and human cognition: the process of individual adaptation to climate change. Glob Environ Chang 15:199–213
61. Bousquet F, Le Page C (2004) Multi-agent simulations and ecosystem management: a review. Ecol Model 176:313–332
62. Magliocca NR, Brown DG, Ellis EC (2013) Exploring agricultural livelihood transitions with an agent-based virtual laboratory: global forces to local decision-making. PLoS One 8: e73241
63. Magliocca NR, Brown DG, Ellis EC (2014) Cross-site comparison of land-use decision-making and its consequences across land systems with a generalized agent-based model. PLoS One 9:e86179
64. Bone C, Dragićević S (2010) Simulation and validation of a reinforcement learning agent-based model for multi-stakeholder forest management. Comput Environ Urban Syst 34:162–174
65. Scheffer M, Bascompte J, Brock WA, Brovkin V, Carpenter SR, Dakos V, Held H, van Nes EH, Rietkerk M, Sugihara G (2009) Early-warning signals for critical transitions. Nature 461:53–59
66. Andersen T, Carstensen J, Hernández-García E, Duarte CM (2009) Ecological thresholds and regime shifts: approaches to identification. Trends Ecol Evol 24:49–57
67. Van Meter KJ, Basu NB (2015) Catchment legacies and time lags: a parsimonious watershed model to predict the effects of legacy storage on nitrogen export. PLoS One 10:e0125971–e0125922
68. Gray SA, Gray S, De Kok JL, Helfgott AER, O'Dwyer B, Jordan R, Nyaki A (2015) Using fuzzy cognitive mapping as a participatory approach to analyze change, preferred states, and perceived resilience of social-ecological systems. Ecol Soc 20:11
69. Davoudi S, Shaw K, Haider LJ, Quinlan AE, Peterson GD, Wilkinson C, Fünfgeld H, McEvoy D, Porter L (2012) Resilience: a bridging concept or a dead end? "reframing" resilience: challenges for planning theory and practice interacting traps: resilience assessment of a pasture management system in northern afghanistan urban resilience: what does it mean in planning practice? Resilience as a useful concept for climate change adaptation? The politics of resilience for planning: a cautionary note. Plan Theory Pract 13:299–333

Deconstructing Geospatial Agent-Based Model: Sensitivity Analysis of Forest Insect Infestation Model

Taylor Anderson and Suzana Dragićević

Abstract Agent-based models (ABM) can be used to represent the spatio-temporal dynamics of real world geospatial phenomena, however because of their complexity, they can be difficult to implement and validate. This study uses the invariant-variant validation approach to further model testing of a developed ABM of forest insect infestation representing spatio-temporal dynamics of the emerald ash borer (EAB). The invariant-variant method deconstructs model results to facilitate an improved understanding of the model's sensitivity to changes in input parameters and focuses on EAB agents' access to information. Obtained results indicate that the developed EAB agent-based model represents and maintains both process accuracy and spatial similarity.

Keywords Complex systems • Agent-based models • Model testing • Invariant-variant method • Emerald ash borer

1 Introduction

Ecological phenomena such as insect infestations can be modelled using a complex systems approach such as cellular automata and agent-based models to better understand how interactions between individuals and their local environment generate spatial patterns at much larger scales [1]. This approach acknowledges that local variation has a significant impact on emergent system behavior. Traditional equation-based ecological models tend to ignore local heterogeneity and model ecological processes from the top-down, limiting their ability to capture system complexity [2]. As an alternative, geospatial agent-based models (ABM) represent the system from the bottom-up, overcoming these limitations. ABMs implement discrete, heterogeneous "agents" to represent real world entities (i.e. an insect) and capture system processes at the local scale. As agents interact with one another

T. Anderson (✉) • S. Dragićević
Spatial Analysis and Modeling Research Laboratory, Department of Geography, Simon Fraser University, 8888 University Drive, Burnaby, BC, Canada V5A 1S6,
e-mail: taylora@sfu.ca; suzanad@sfu.ca

L. Perez et al. (eds.), *Agent-Based Models and Complexity Science in the Age of Geospatial Big Data*, Advances in Geographic Information Science,
DOI 10.1007/978-3-319-65993-0_3

and their virtual environment over time, complex system level behavior and spatial patterns emerge. Furthermore, ABMs can be integrated with geographic information systems (GIS), facilitating the representation of the environment in which the agents interact using real geospatial data.

It has been demonstrated that geospatial ABMs can capture the complexity of the real-world systems and have been used to accurately represent ecological phenomena such as fish [3], birds [4], and forest insect infestations such as the mountain pine beetle [5, 6] and the emerald ash borer [7, 8]. ABMs provide a useful methodology for the evaluation of future policy decisions and actions, sometimes referred to as scenario planning [9]. For example, using an ABM as a virtual laboratory, Anderson & Dragicevic [8] develop scenarios to explore and optimize the biological control of the EAB forest insect infestation i.e. determine how many biological control agents need to be released and where they need to be released to be effective.

To use an ABM in the decision-making process, the level of confidence of the model to represent the phenomena realistically must be demonstrated. However, building and implementing an ABM capable of capturing the complexity of real world geospatial phenomena presents unique challenges in both understanding and communicating their validity. Particularly, as ABMs represent behavior of various agents, they rely on stochasticity, and thus may produce a variety of results, even when using the same input parameters [10]. This can make testing using traditional map comparison techniques and accuracy assessments that measure spatial similarity between model outputs and reference data difficult, as these measures may hide or ignore these important variations [11].

For example, the variation in results may be a function of path dependence, where positive and negative feedback processes have driven the model produce two or more distinct spatial patterns across model runs. The patterns that emerge from these processes may fluctuate between matching the patterns found in reference data and vice versa. However, as a bottom-up modelling methodology, ABMs seek to represent the underlying dynamics and processes in producing complex system level behavior and thus their usefulness may not be fully measured through aggregate pattern matching. Thus, it may be valuable to also explore the model's process accuracy and increase confidence that the model can represent the *processes* driving the spatial patterns of the phenomena. Additionally, small changes in ABM input parameters may generate disproportionally large variations in output spatial patterns. Understanding how model input parameters affect model outputs is an important step in developing functional and useful ABMs [12].

The invariant-variant method developed by Brown et al. [13] makes the distinction between model results that remain consistent across model runs (invariant) and model results that change across model runs (variant). The deconstruction of model results into these two classes is useful in the identification of the underlying model processes that give rise to emergent spatial patterns. Furthermore, the invariant-variant method and can aid in sensitivity analysis to clearly understand how changes in input parameters change model results. These methods have been advanced to account for not only spatial variation, but also temporal variation across model runs [14], where Bone et al. develop a temporal invariant-variant approach to account for transition between land use classes over time.

The purpose of this study is to further the model testing of a forest insect infestation geospatial ABM developed by Anderson & Dragicevic [7, 8] using the invariant-variant method. The developed ABM simulates emerald ash borer (EAB) forest insect infestation dynamics and spread in Oakville, Ontario, Canada for 2 years (2008–2009). Geospatial data delineating real EAB extent in 2009 obtained from Oakville facilitates model testing using this approach. The main objective of this study is to deconstruct and better understand model results using the invariant-variant method and to test the sensitivity of the model parameters. The following sections will provide a brief outline of the developed EAB ABM and present the model testing method and results, finishing with a discussion and conclusions.

2 Background

2.1 *Emerald Ash Borer (EAB)*

The emerald ash borer (EAB) is an invasive bark beetle, native to countries in Asia [15]. The beetle was thought to be introduced into North America in the late 1990s and was discovered in 2002 in Detroit, Michigan, USA. Since its introduction into the region, the pest has been responsible for the decline of the North American ash tree population, creating devastating ecological and economic impacts. Eradication has been unsuccessful due to challenges in infestation detection, a lack of native predators, and long-distance dispersal patterns that are difficult to predict [16].

EAB complete their lifecycle in one (sometimes two) years and consists of the stages: active larvae, inactive larvae, pupae, and adulthood [17]. The EAB eggs mature into larvae and then into adulthood while under the bark of ash trees, a process that takes almost 1 year, before emerging in early June through August, with peak emergence in mid-July [18]. The beetle uses olfactory and visual cues to determine the most suitable hosts and prefer to lay their eggs in ash trees that are stressed [19], have a lower natural resistance to insect infestations such as green, black, and white ash [20], and are larger in size and capable of supporting the larval galleries [21]. EAB find their hosts through local dispersal, travelling on average 2.8 km/day [22]. The beetles spread can be exacerbated by long-distance dispersal, facilitated by the movement of infested saplings or firewood. These two dispersal mechanisms generate a pattern called stratified dispersal, where eventually the natural front of infestation and satellite populations coalesce [23].

2.2 *EAB ABM*

It is important to understand the spatial patterns and processes of insects' dispersal, interactions, and dynamics, but this information can be difficult to obtain from field measurements. Existing EAB models use differential equations [24] and diffusion

Table 1 State variables and parameters of EAB adult and EAB larvae

State variables and parameters of EAB		
Variable	Description	
ID	The agent's unique identifier	
Age	The agent's age	
Geography	The location (decimal degrees) of the agent	
EAB adult agent parameters		
Parameter	Description	EAB value
Maximum flight distance/day	Flight mill tethering distance that females can travel/day	2.8 km/day [22]
Chance of fertility	Average fertility rate of females	82% [27]
Maximum number of offspring	Average threshold for maximum offspring	Randomly selected value between 60 and 90 offspring/individual [28]
Survival rate of eggs	Survival rate as a function of chance	Randomly selected between and 53–65% survive [28]
EAB larvae agent parameters		
Sex ratio	Female: Male	1:1, 50% [18]
Survival rate of larvae	Survival rate as a function of tree resistance, disease, and predation via other species i.e. woodpecker	Host tree defense: max 21.5% Disease: 3% Woodpecker: max 17% [29]

models [25], however are limited in their representation of complexity inherent to insect infestation processes and behavior of the beetles [26]. Alternatively, ABMs can be used to simulate these processes and better understand complexity of the infestation dynamics and use scenarios to aid in management and decision making.

Anderson and Dragicevic [7] have proposed an EAB ABM to simulate spatio-temporal dynamics of EAB in Oakville, Ontario, Canada for 2 years (2008–2009), and was further enhanced [8] to explicitly represent EAB population dynamics. The model is composed of agents that represent individual EAB in larvae and adult stages. Agents are programmed with state variables and parameters that are unique to each individual (Table 1). State variables track the state of an agent at each iteration such as age and location. Parameters characterize an individual agent's biological properties such as the chance of fertility and the maximum number of offspring an individual may produce. These parameters are determined using biological information documented in EAB literature. Agent behavior is driven using several subroutines that execute stages in the life cycle including local dispersal, long-distance dispersal, mating and fertility, maturity, infestation of ash trees, and death (Table 2).

The model simulates EAB spatio-temporal dynamics over a period of two seasons of EAB infestation from June 1st, 2008 (T_1) when the EAB was first introduced to the region, to the end of August 2009 (T_{460}) [8]. Each iteration in the model (T_i) represents 1 day $i = (1, 460)$ in the real world. Due to random processes in the model, no two simulation outputs are the same. Therefore, the model is executed 50 times to generate a statistically significant distribution of results, where each run generates a variation of the emergent patterns of EAB infestation in 2009.

Table 2 Subroutines that generate agent behavior

Agent processes	
EAB adult agent	
Process	Description
Aging	The age of each agent is increased by 1 day at each new iteration. The age (in days) of an agent triggers the execution of life cycle processes
Short-distance dispersal	Short distance dispersal is the process whereby agents change their location. Short distance dispersal begins after EAB adults emerge at the age of 1 day and continues throughout the rest of the agent's lifetime. The distance in which an individual EAB agent will move at each model iteration is a function of: (1) the average distance EAB travel per day (2.8 km) [22] and (2) host suitability [30]
	The flight distance of 2.8 km per day bounds the EAB agents' access to information about their environment (i.e. what trees are available). Each EAB may search within a radius of their average daily flight distance for host trees and compare them with one another based on their suitability. The comparison between trees by EAB is controlled by a host selection algorithm, developed by Anderson & Dragicevic [7] that allows EAB agents to optimize their decision of which tree to infest based on their preferences. EAB host selection preferences have been studied extensively and are a function of (1) tree distance, (2) tree type, (3) tree stress, and (4) tree size. Specifically, EAB prefer trees which are closer in distance, tree types of lower resistance to infestation such as the green ash, trees which are under stress perhaps due to existing infestation or age, and trees larger in size
Mate	EAB agents may become fertile based on their chance of fertility. Those that become fertile, mate at the age of 7 days. EAB are randomly assigned a maximum number of offspring between 60 and 90 individuals [31]
Oviposit	EAB agents become fully mature and begin seeking suitable ash trees using the host selection algorithm to host their larval galleries at age 10 days. At each iteration, EAB oviposit a random number of eggs onto their choice of tree. This process continues until the maximum number of offspring have been produced. The number of eggs may be reduced based on their chance of survival
Death	EAB agents die once they have produced their maximum number of offspring
Long-distance dispersal	Long distance dispersal is a random process in the model where satellite populations (sometimes 1% of the original population) becomes established in regions of high susceptibility to this process i.e. along major transportation networks or near campgrounds. The environment in which the EAB interact is representative of Oakville's urban forest and is based on Oakville's tree inventory geospatial data sets
EAB larvae agent	
Death	Larvae may die as a result of tree resistance, disease, and native predators [29]. This process uses a random number generator to determine how susceptible the larvae is to these factors
Emergence	EAB larvae emerge when they reach the age of 340 days and if it is female. A random number generator is used to determine the sex of the larvae

3 Methods

Initial model testing of the EAB ABM has been performed. The model has been calibrated to simulate the real-world rate of spread, determined by using real world data delineating the extent of EAB infestation from 2002–2010 [7, 8]. Specifically, the model has a simulated rate of spread from the epicenter of infestation in 2008 to the delineation of EAB infestation in 2009 of 2.119 km/year in comparison to the observed rate of spread in reality of 2.098 km/year. Additionally, the model simulates spread with an average distance of 4238.77 m and a maximum distance of 11049.50 m in comparison to the observed average distance of 4196.17 m with a maximum distance of 11186.3 m [7].

Although research has shown tree type, tree size, tree stress, and tree distance are the driving factors in host selection and are included in the host selection algorithm, the order in which EAB prioritize these factors is unknown. Therefore, Anderson & Dragicevic [8] performed the sensitivity analysis to determine the sensitivity to the order in which these factors are preferred i.e. whether EAB prefer trees that are closer or are more stressed. Initial model validation used traditional methods of map comparison between model outputs and real-world data and included the following metrics: (1) the spatial agreement between the model output and the real-world data in location of infestation in 2009 and (2) the spatial agreement between the model output and the real-world data in severity of infestation. The level of agreement of the state of the trees between model outputs and real-world data was determined. The overall accuracy of the model calculated by using these methods was found to be 72% in simulating the location of EAB infestation [7, 8] and 64% overall accuracy in forecasting location of severity of infestation [7]. Although a useful starting point for evaluating the overall performance of the model, simple accuracy assessments using map comparison techniques may not allow for in depth exploration of the model processes that may be contributing to the distribution of model results. Therefore, the invariant-variant method is used to further the EAB ABM model testing and sensitivity analysis.

3.1 Invariant-Variant Method for Analysis of EAB ABM

In the case of the EAB ABM, the *invariant region* can be defined as the trees that are always or almost always infested or always or almost always not infested and the *variant region* can be defined as the trees that are sometimes infested and sometimes not infested. To determine which trees are invariant or variant across model runs, the EAB ABM was run 50 times, producing a statistically significant distribution of results. Each run of the EAB ABM outputs a geospatial dataset containing all trees and their corresponding attributes (i.e. tree height, tree DBH) and infestation status (i.e. whether the tree has been simulated as infested or not). The infestation status of a tree across all model runs is used to calculate the proportion of runs in which the tree is infested, denoted as t_{xy} at location x,y. For example, if tree t is infested in 46 of a possible 50 runs, $t_{xy} = 0.92$, meaning that the tree is infested in 92% of the model runs.

The invariant and variant trees are partitioned using a threshold θ. For example, trees that are invariant and infested *ID* are defined by a threshold $\theta = 0.9$, as used by Brown et al. (2005), and as such must be infested in at least 90% of model runs. Therefore, *ID* is the number of trees $t_{xy} > \theta$. The *ID* region is compared with the real-world data delineating EAB infestation in 2009 and sub-classified into invariant correct *IC* and invariant incorrect *II*. *IC* are trees that are infested in 90% of model runs and infested in reality. Conversely, *II* are trees that are infested in 90% of model runs and are not infested in reality. Because these trees are invariant, every model run will have nearly the same value for *IC* and *II*. In contrast to *ID*, trees that are rarely infested, $t_{xy} < 1\text{–}\theta$, are denoted as *IU*, meaning they are infested in less than 10% of model runs.

Trees that are *variant* are sometimes simulated as being infested (11–89% of model runs). In addition to trees that are correctly simulated as infested in the *invariant* region *IC*, trees may be correctly simulated as infested in the *variant* region. The number of variant correct *VC* is a function of a particular run k. If C_k is used to denote the number of infested locations that are predicted by a single run k, then $C_k = IC + VC_k$. *VC* can be plotted using a histogram to show model behavior across all of the runs. A histogram that has a set of runs with extremely high *VC* and a set of runs with low *VC* may indicate multiple *paths.*

Decomposing model results into its invariant and variant regions allows for the identification of patterns that may not be obvious when looking at the overall generated spatial patterns of infestation. A small *IC* and a large *VC* may indicate that the model is path dependent, where complex dynamics of the phenomena represented by the ABM causes the generation of multiple spatial patterns. For example, in some runs infestation spreads to unexpected locations and in others, infestation coincides with the reference data. This is important, because when calculating a simple accuracy assessment, a model that produces this variation in results may not be within acceptable limits of accuracy, however the model may be path dependent, evidence of the model's ability to capture system processes accurately. In contrast, if *IC* is large and *VC* is small on average, it can be concluded that the accuracy of the developed EAB ABM model primarily originates from getting the large invariant region correct.

3.2 Bounded Rationality Sensitivity Tests

The sensitivity of the model to the EAB agent's access to information was tested. To test the impact that an increase in EAB access to information would have on the model simulation outcomes, the model was run 50 times with an increased flight distance of 5.6 km per day, double that of the original distance. Furthermore, the impact that a decrease in the EAB agents' access to information on the model simulation outcomes was tested using a flight distance of 1.4 km/day, half of the original distance, and was run 50 times.

4 Results

4.1 *Invariant-Variant Method for Analysis of EAB ABM*

The simulation results obtained by the invariant-variant analysis for the EAB ABM are presented in Table 3A and Fig. 1a. The EAB ABM model generates a high *IC* (invariant infested correct) at 1419 trees and a high *IUC* (invariant uninfested correct) at 2089 trees versus a low *VC* (variant correct) at 926 trees, meaning that the models map comparison accuracy primarily comes from getting the invariant region correct.

The invariant region, where infestation occurs in over 90% of model runs, is located near the center of the study area, the core zone, where EAB first were identified in this region in the real-world (Fig. 1a). The simulated invariant region

Table 3 Invariant-variant analysis results for sensitivity of EAB agents' (A) access to information using a flight distance of 2.8 km/day, (B) reduced access to information using a distance of 1.4 km/day, and (C) increased access to information using a distance of 5.6 km/day

Distance parameter	Description	(A) 2.8 km/day	(B) 1.4 km/day	(C) 5.6 km/day
Invariant infested (ID)	Simulated as infested in 90% or more of model runs	1619	727	1912
Invariant correct (IC)	Simulated as infested in 90% or more of model runs and is in agreement with the reference data	1464	724	1643
Invariant incorrect (II)	Simulated as infested in 90% or more of model runs and is not in agreement with the reference data	155	3	269
Invariant uninfested (IU)	Simulated as uninfested in 90% or more of model runs	3355	4829	1904
Invariant ucorrect (IUC)	Simulated as uninfested in 90% or more of model runs and is in agreement with the reference data	2089	2445	1242
Invariant uincorrect (IUI)	Simulated as uninfested in 90% or more of model runs and is not in agreement with the reference data	1266	2384	662
Variant (V)	Sometimes simulated as infested	1208	626	2336
Variant correct (VC)	Sometimes simulated as infested and is infested in reality	926	548	1351
Variant incorrect (VI)	Sometimes simulated as infested and is not infested in reality	282	78	1015

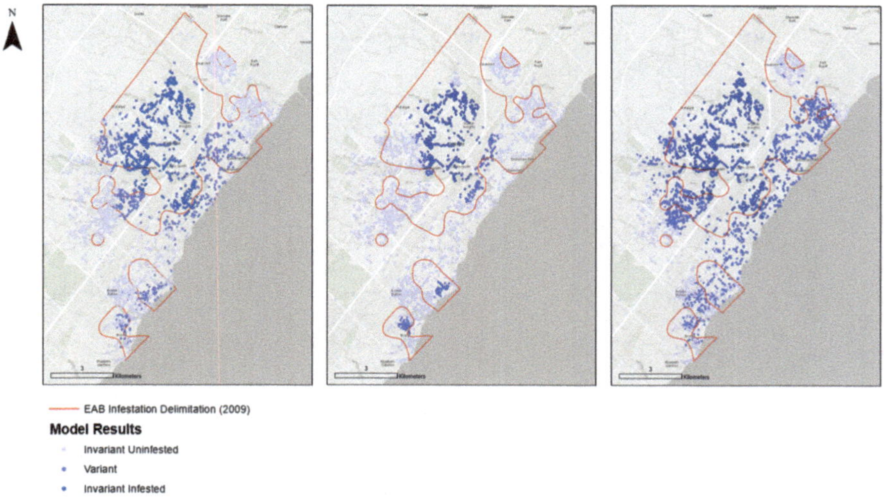

Fig. 1 Locations of variant and invariant trees based on simulations incorporating EAB agents' with (**a**) access to information using a flight distance of 2.8 km/day, (**b**) reduced access to information using a distance of 1.4 km/day, and (**c**) increased access to information using a distance of 5.6 km/day

mostly falls within the delineation of EAB infestation obtained from real-world data. The variant region, where infestation occurs in some runs and not in others, is located on the perimeter of this core invariant zone and in satellite population zones. In addition, there are a few variant infested trees that fall between the core zone and the satellite population zones.

As presented in Fig. 1a, the model underestimates the number of infested trees, meaning that 1266 invariant uninfested trees are infested in the real world. The model does well at predicting the number of invariant infested trees and rarely does the model predict a tree is infested when it is not infested in reality. The distribution of model runs *k* and the number of trees accurately simulated as infested is presented in Fig. 2. The histogram depicts the variance across model runs.

4.2 Bounded Rationality Sensitivity Tests

The results indicate that the EAB ABM is sensitive to the EAB agents' access to information. Specifically, as presented in Fig. 1b, reducing the EAB agents' access to information affects the simulated outcomes in the following ways: (1) the invariant infested region is smaller, but more accurate; (2) the invariant uninfested region becomes much larger, but becomes much less accurate; and (3) the variant region becomes smaller, but more accurate. In general, reducing the EAB agents' access to information underestimates the number of trees that are infested in the real-world by almost double of that of the original model at 2384 trees (Table 3B).

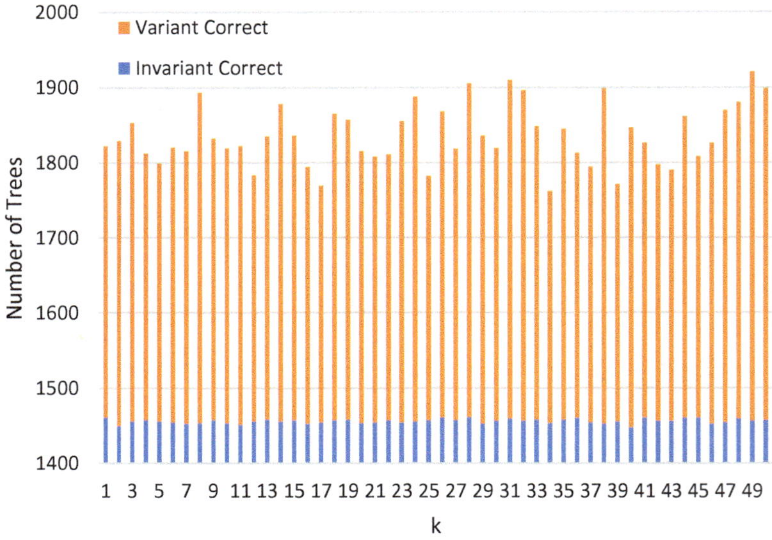

Fig. 2 Variability of number of trees correctly simulated as infested across 50 model runs including invariant correct and variant correct

Reducing the EAB agents' access to information maintains the emergence of the invariant region located in the core zone. Variant regions emerge around the perimeter of this core zone and in satellite population zones. All trees infested in the simulation in this scenario fall within the real-world delineation of the EAB infestation (Fig. 1b). Reducing the EAB agents' access to information eliminates the variant region between the two zones.

In contrast, as presented in Fig. 1c, increasing the EAB agents' access to information affects the simulated outcomes as such: (1) the invariant infested region is slightly larger, with similar accuracy to the original model; (2) the invariant uninfested region is much smaller, but does not overestimate uninfested trees; (3) the variant region becomes much larger, larger than the invariant infested region, but overestimates infestation in trees that are not infested in reality at 1015 trees (Table 3C).

Increasing the EAB agents' access to information maintains the generation of the invariant region located in the core zone and the invariant region around the perimeter of the core zone and in satellite zones and increases the variant region that falls between these two zones (Fig. 1c).

5 Discussion and Conclusions

The variable distribution of the frequency of trees correctly predicted as infested across model runs (Fig. 2) may indicate that EAB ABM generates multiple paths. A primary assumption would be that the stochastic long-distance dispersal processes are generating the variation in accuracy from model run to model run. In the simulation outputs, small satellite populations sometimes appear in the south-west part of the study site due to the location's proximity to the highway and the Bronte Creek Provincial Park and because long distance dispersal is a random process in the model, simulated satellite populations are always variant. There is a slight positive relationship ($R^2 = 0.38$) between the model's overall accuracy and the accuracy in forecasting satellite populations, meaning that model runs that predict satellite populations are sometimes more accurate and thus may explain some of the variability in model runs. Long distance dispersal is not often spatially similar to the locations of reference data, which would reduce the accuracy of the model when using traditional map comparison and accuracy assessments. However, long distance dispersal may be variant correct, indicating process accuracy.

The invariant-variant analysis demonstrates that the model is sensitive to reducing the EAB agents' access to information. Reducing the flight distance to 1.4 km/day results in a severe underestimation of the number of trees infested in reality. This is evident by the decrease in the invariant infested region and the increase in the invariant uninfested region (Table 2; Fig. 1b). In contrast, the results suggest that the model is less sensitive to an increase in the EAB agents' access to information with a flight distance of 5.6 km/day. Specifically, the invariant infested region and the invariant uninfested region are similar, if not more accurate than the original model parameter of 2.8 km/day (Table 2). This can be attributed to the host selection algorithm which acts as a negative feedback mechanism by prioritizing the infestation of trees that are closer in distance and thus accurately simulates infestation processes. However, with an increase in access to information, the variant region increases substantially (Fig. 1c) which means that in some model runs, EAB infestation is overestimated.

Real-world EAB infestation at regional scales undergo the process of stratified dispersal, where the core zone and satellite population zones merge, advancing the front of EAB spread at increased rates. Evidence of the stratified dispersal process can be identified in the simulations, where the core zone and satellite zones begin to merge in some simulation runs, thus developing a variant infested region between the two. Specifically, the early stages of a merge between infestation in the core zone and satellite population zones occurs in some runs of the original model and is even more pronounced when the EAB agents' access to information is increased. In the reference data, however, the two zones including the core zone and the satellite population zones are entirely separate. Thus, traditional accuracy assessments and map comparisons would deem model runs that simulate stratified dispersal as inaccurate and ignore the value in the model's ability to simulate this important process.

In summary, ABM testing can be a challenging process. Common spatial model evaluation measures such as map comparison or other simple accuracy assessments are difficult to apply since ABMs produce a variable distribution of outputs across model runs in response to agents' individual behavior and interactions in combination with stochasticity, local heterogeneity, feedbacks, and evolution in the model [6]. Using these conventional measures may provide an understanding of the spatial similarity between aggregate spatial patterns in the reference data and aggregate spatial patterns generated as model outputs. This can provide initial confidence in model performance. The invariant-variant analysis breaks down the aggregate measure of spatial similarly and provides insight as to what may be influencing these measures, thus improving the understanding of the model processes that generate model results and help the modeler gain confidence that the real-world phenomena is represented realistically.

EAB infestation poses significant threats to forest ecosystems across Canada and in the US. The developed EAB ABM can be used to aid in meeting management goals by evaluating how various management actions impact infestation dynamics. However, naturally, before the results can be used to make decisions, sufficient data demonstrating that the model's results are valid must be attained. The invariant-variant analysis demonstrates the proposed agent-based model possesses the ability to represent underlying processes driving emergent patterns of EAB spread to assist and give confidence to decision makers such as stakeholders or policy makers in model outputs and reduce the possibility of making unsuitable decisions and risk time and money. In particular, the variant and more unpredictable nature of satellite populations may require a focus of resources by decision makers in order to slow the infestation front and reduce large scale negative impacts of EAB infestations.

Acknowledgments This study was fully funded by a Natural Sciences and Engineering Research Council (NSERC) Canadian Graduate Scholarship-Doctoral (CGS D) awarded to the first author and Discovery Grant awarded to the second author. The datasets were provided by the Town of Oakville, Canada. The authors are thankful to Compute Canada WestGrid high-performance computing facility for enabling model simulations.

References

1. DeAngelis DL, Mooij WM (2005) Individual-based modeling of ecological and evolutionary processes. Annu Rev. Ecol Evol Syst 36:147–168
2. Grimm V, Railsback SF (2005) Individual-based modeling and ecology. Princeton University Press, New Jersey
3. Letcher BH, Rice JA, Crowder LB, Rose KA (1996) Variability in survival of larval fish: disentangling components with a generalized individual-based model. Can J Fish Aquat Sci 53(4):787–801
4. Travis JM, Dytham C (1998) The evolution of dispersal in a metapopulation: a spatially explicit, individual-based model. Proc R Soc Lond B Biol Sci 265(1390):17–23
5. Pérez L, Dragićević S, White R (2013) Model testing and assessment: perspectives from a swarm intelligence, agent-based model of forest insect infestations. Comput Environ Urban Syst 39:121–135
6. Bone C, Altaweel M (2014) Modeling micro-scale ecological processes and emergent patterns of mountain pine beetle epidemics. Ecol Model 289:45–58
7. Anderson T, Dragićević S (2015) An agent-based modeling approach to represent infestation dynamics of the emerald ash borer beetle. Ecol Inform 30:97–109
8. Anderson T, Dragicevic S (2016) Geospatial pest-parasitoid agent based model for optimizing biological control of forest insect infestation. Ecol Model 337:310–329

9. McLane AJ, Semeniuk C, McDermid GJ, Marceau DJ (2011) The role of agent-based models in wildlife ecology and management. Ecol Model 222(8):1544–1556
10. Wilensky U, Rand W (2015) An introduction to agent-based modeling: modeling natural, social, and engineered complex systems with NetLogo. MIT Press, Massachusetts
11. Pontius RG (2000) Quantification error versus location error in comparison of categorical maps. Photogramm Eng Remote Sens 66:1011–1016
12. Ligmann-Zielinska A, Sun L (2010) Applying time-dependent variance-based global sensitivity analysis to represent the dynamics of an agent-based model of land use change. Int J Geogr Inf Sci 24(12):1829–1850
13. Brown DG, Page S, Riolo R, Zellner M, Rand W (2005) Path dependence and the validation of agent-based spatial models of land use. Int J Geogr Inf Sci 19(2):153–174
14. Bone C, Johnson B, Nielsen-Pincus M, Sproles E, Bolte J (2014) A temporal variant-invariant validation approach for agent-based models of landscape dynamics. Trans GIS 18(2):161–182
15. Straw NA, Williams DT, Kulinich O, Gninenko YI (2013) Distribution, impact and rate of spread of emerald ash borer Agrilus Planipennis (Coleoptera: Buprestidae) in the Moscow region of Russia. Forestry 86(5):515–522
16. Fahrner SJ, Lelito JP, Blaedow K, Heimpel GE, Aukema BH (2014) Factors affecting the flight capacity of tetrastichus planipennisi (hymenoptera: Eulophidae), a classical biological control agent of Agrilus Planipennis (Coleoptera: Buprestidae). Environ Entomol 43(6):1603–1612
17. Cappaert D, McCullough DG, Poland TM, Siegert NW (2005) Emerald ash borer in North America: a research and regulatory challenge. Am Entomol 51(3):152–165
18. Lyons DB, Jones GC (2005) The biology and phenology of the emerald ash borer. In: Proceedings, 16th US Department of Agriculture interagency research forum on gypsy moth and other invasive species, pp 62–63
19. McCullough DG, Poland TM, Anulewicz AC, Cullough DGMC (2009) Emerald ash borer (Coleoptera: Buprestidae) attraction to stressed or baited ash trees. Environ Entomol 38(6):1668–1679
20. Rebek EJ, Herms DA, Smitley DR (2008) Interspecific variation in resistance to emerald ash borer (Coleoptera: Buprestidae) among north American and Asian ash (Fraxinus spp.) Environ Entomol 37(1):242–246
21. Mercader RJ, Siegert NW, Liebhold AM, McCullough DG (2011) Influence of foraging behavior and host spatial distribution on the localized spread of the emerald ash borer, Agrilus Planipennis. Popul Ecol 53(2):271–285
22. Taylor RA, Poland TM, Bauer LS, Windell KN, Kautz JL (2007) Emerald ash borer flight estimates revised. In: Proceedings of the emerald ash borer/Asian longhorned beetle research and technology. FHTET-2007-04, US Department of Agriculture Forest Service, Forest Health Technology Enterprise Team, Morgantown, West Virginia
23. Siegert NW, McCullough DG, Liebhold AM, Telewski FW (2008) Dendrochronological reconstruction of the establishment and spread of emerald ash borer. In: Mastro V, Lance D, Reardon R, Parra G (eds). In: Proceedings, the emerald ash borer and asian longhorned beetle research and technology development meeting. Morgantown, West Virginia
24. Barlow LA, Cecile J, Bauch CT, Anand M (2014) Modelling interactions between forest pest invasions and human decisions regarding firewood transport restrictions. PLoS One 9(4):e90511
25. Muirhead JR, Leung B, Overdijk C, Kelly DW, Nandakumar K, Marchant KR, MacIsaac HJ (2006) Modelling local and long-distance dispersal of invasive emerald ash borer Agrilus Planipennis (Coleoptera) in North America. Divers Distrib 12(1):71–79
26. With KA (2002) The landscape ecology of invasive spread. Conserv Biol 16(5):1192–1203
27. Rutledge CE, Keena MA (2012) Mating frequency and fecundity in the emerald ash borer Agrilus Planipennis (Coleoptera: Buprestidae). Ann Entomol Soc Am 105(1):66–72
28. Jennings DE, Taylor PB, Duan JJ (2014) The mating and oviposition behavior of the invasive emerald ash borer (Agrilus Planipennis), with reference to the influence of host tree condition. J Pest Sci 87(1):71–78

29. Duan JJ, Ulyshen MD, Bauer LS, Gould J, Van Driesche R (2010) Measuring the impact of biotic factors on populations of immature emerald ash borers (Coleoptera: Buprestidae). Environ Entomol 39(5):1513–1522
30. MacFarlane DW, Meyer SP (2005) Characteristics and distribution of potential ash tree hosts for emerald ash borer. For Ecol Manag 213(1):15–24
31. BenDor TK, Metcalf SS, Fontenot LE, Sangunett B, Hannon B (2006) Modeling the spread of the emerald ash borer. Ecol Model 197(1–2):221–236

An Agent-Based Model to Identify Migration Pathways of Refugees: The Case of Syria

Guillaume Arnoux Hébert, Liliana Perez, and Saeed Harati

Abstract The Syrian civil war has generated a refugee crisis in the Middle East and Europe. This study draws on complex systems theory and the agent-based modelling method to simulate the movement of refugees in order to identify pathways of forced migration under the present crisis. The model generates refugees as agents and lets them leave conflict areas for a destination that they choose based on their respective characteristics and desires. The simulation outputs are compared with existing data regarding the state of forced migrations of Syrians to assess the performance of the model.

Keywords Conflict induced migration • Syrian refugees • Agent-based modeling • GIS • Migration pathways

1 Introduction

Survival and wellbeing are two important characteristics of human nature. As witnessed through history with the repeated population displacements [1, 2], people who find their social conditions dissatisfactory will often migrate to places that promise better possibilities for improvement; for example, peace and wealth compared to the violence and despair that characterize their home countries. The reasons for migration vary from economic, political and security causes, to natural or anthropogenic disasters. Nevertheless the hope is always the same, to find a safe and better place to live and prosper [1]. This is also the case for many Syrians who have faced insecurity and despair brought on by a civil war that been going on for several years now [3–6].

G. Arnoux Hébert • L. Perez (✉) • S. Harati
Laboratoire de Géosimulation Environnementale, Department of Geography, Université de Montréal, Pavillon 520 Côte-Sainte-Catherine, Montréal, QC, Canada H3T 1J4,
e-mail: guillaume.arnoux.hebert@umontreal.ca; l.perez@umontreal.ca; saeed.harati.asl@umontreal.ca

L. Perez et al. (eds.), *Agent-Based Models and Complexity Science in the Age of Geospatial Big Data*, Advances in Geographic Information Science,
DOI 10.1007/978-3-319-65993-0_4

A multitude of armed groups, which form an intricate web of alliances characterised by varied sets of commonalities and contradictions, add a layer of complexity to the situation. In addition, global and regional powers are involved in the conflict sometimes fighting among themselves to pursue opposite objectives [6]. In that context, it is understandable that much of the population would like to flee the region, triggering the beginning of a great migration. To better understand the dynamics behind migration, this paper presents a model that simulates the pathways the migrants use to flee their home country.

This study uses a dynamic approach to model the decision steps of the migrants, namely, the decision to leave, choices of destination and pathway as well as the decision to stay at the destination. The methodology presented here is rooted in the science of complexity and uses the agent-based modelling (ABM) approach to simulate the dynamics of migration.

The present chapter is organized in five sections. Section 2 provides, an overview on forced migration, complex systems modeling approaches and how these methods have been used to study population migration processes. Input data and the model are detailed in Sect. 3. Section 4 addresses the results, and Sect. 5 concludes the chapter.

2 Review of Literature

2.1 Conflict-Induced Displacement: Contextualizing Refugee Migration

As stated by Zetter [7], terminology is essential when addressing population displacement, and the use of incorrect terminology could have dreadful consequences on the reception and the treatment of a displaced population when reaching a safer or more desirable destination. Language misuse can be used by some countries to hide the reality of conflict-induced displacement and therefore deny the population counts of those forced to flee and find refuge [7]. By definition all refugees are migrants, however, Schmeild [8] stated that until 1990 the study of migration phenomenon was a distinct field from the study of refugee migration, which was considered to be a political phenomenon and as a consequence, was ignored by most migration literature and studies. Since then, the approach to study refugee displacement has changed and refugees' behaviour has been studied as part of migration phenomenon [9]. In general, migration refers to the permanent movement of people to a new area or country [8], while population displacement makes reference to people's movement but only within a specific time frame and generally inside a country [10, 11]. For this study, we have defined refugee migration as the event that occurs when people are forced to flee their homes as a result of a civil war. Likewise, we have adopted the legal definition of a refugee provided in 1951 by the United Nations Convention Relating to the Status of Refugees. The Convention defines a refugee as a person residing outside his or her country of nationality, who is unable or unwilling to return because of a well-founded fear of persecution on account of race, religion, nationality, membership in a political social group, or political opinion [12].

Within the literature, the study of migration is generally classified into distinct areas based on the reasons for population migration. Amongst the most studied causes for migration are economic reasons [13], climatic reasons [14–17], and conflict induced reasons [9, 11, 17]. Another important aspect considered when studying migration is the scale at which the process of migration is examined. Such as scale of examination and analysis could be national, regional or continental scale.

2.2 *Complex Systems Theory*

The conceptual framework of complex systems theory focuses on the many characteristic behaviours of dynamic systems such as self-organization, emergence, non-linearity, path dependence, bifurcation and sensitivity to initial conditions, amongst others [18, 19]. The modelling approaches that draw on complex systems theory, can be used to investigate how the interactions between parts can create collective behaviour within a dynamic system [20, 21] such as human migration. With the objective of modelling and examining the laws governing the behaviour of complex geographic phenomena such as forced migration, agent-based modelling (ABM) approach can be used to study the spatial patterns resulting from the complexities of human migration.

As it is the case with all complex system models, when studying human forced migration, it is important to understand the elements that characterise this phenomenon. In the case of forced migration in Syria, the movement of the population as well as the uniqueness of each individual, in addition to every aspect of the conflict creating the forced migration, are important considerations. Some research has been done in with these considerations [22, 23], but at a very local scale. The decision-making process with autonomous individuals in a bounded-rational environment, such as that of a refugee migration, is in nature heterogeneous and lends itself well to ABM as a tool for analysis [24]. Individuals represented by agents are dynamically interacting with other agents based on simple rules that will give rise to complex behaviours and patterns of displacement.

2.3 *ABM and Forced Migration*

The study of forced migration using ABM is still in its early stages, with statistical modelling still dominating the field [8, 17, 25, 26]. To understand the migration of Syrian refugees, most researchers have used static and statistical approaches to count the number of migrants leaving, those in transit and others arriving in each country [5, 27–29]; however, most of the studies done on forced migration have been linked to climate change [10, 16, 17, 30]. In addition, there are a few studies on conflict induced migration which have been conducted by people with an expertise in political science or sociology [11, 31], and not in geography (i.e. spatial dynamic modelling). The lack of research on migration patterns modelling can be associated with two main challenges related to the choice of decision rules and the use of empirical data [32]. Decision rules comprise the part of the model used to replicate the decision making process of a human being. Defining realistic decision rules can be difficult and that is one of the reasons why there are not many dynamic models on conflict induced migration. In this study we use a Psycho-Social and Cognitive approach [32], that is based on the planned behaviour theory [33]. Planned behaviour theory states that an individual that processes information, mediates the effects of biological and environmental factors on one's behavior. Thus, whether a behavior, for example migration, occurs or not is the result of the probability that the influence factors are compelling enough for each individual. The advantage of the Psycho-Social and Cognitive model is that it allows the inclusion of an infinite number of features to model decision making process as well as takes into account social influence and the uncertainty of life [33]. Empirical data constitutes one very important part of model development; without it, it is very difficult to parameterize and calibrate a realistic model [32]. Due to strategic reasons, valid empirical data within a conflict zone are hard to obtain, and this is why there are not many models on conflict induced migration.

When creating a migration model there are three major aspects to think about. The first is to know the moment the populations decide to leave (When). The second is to know the destination of the migrants (Where) and finally to know whether they want to stay or not at their destination. These aspects identify migration pathways.

The first aspect or the moment people decide to leave is the one that is the most studied and conceptualised [11, 14]. The choice to leave is, according to Oliver-Smith [30], not a reaction to an event, but it is due to an accumulation of factors that make people leave. Sokolowski and Banks [11] agree and add that people in risk zones choose the best options based on an analysis of circumstances, risks, and cost and benefit. That is what Klabunde and Willekens [32] present as part of planned behavior theory.

The second aspect or the choice of destination is a complex problem with multiple components; Moore and Shellman [34] investigate if refugees are more likely to relocate themselves inside or outside their country, while Schmeild [8] affirm that refugees usually go to neighboring countries with the same ethnic group and religion. In the case of Syria, refugees do not want to stay in the neighboring countries, instead they prefer to go to a country with different culture and religion (mostly in Europe) because they perceive that life in these countries is better than the ones in the Middle East [4]. Although refugees do not see all European countries on the same scale, some of them may try to enter one of the Schengen nations with the perspective to move to another if the standard of living is better [35]. In general, literature states that migrants/refugees prefer cities or countries that are politically stable, richer and safer than their departing location. It is also reported that another important aspect within the decision-making process is the capacity of absorption of migrants at each destination and the reception attitude of the host population [3–5]. Likewise, refugees will been keen to move to a rich or a more prosperous country but they will also want to move to a country in which they have family ties [36].

The third aspect is related to the degree of satisfaction with the chosen destination, and it can be divided into two different situations. The first one reports on migrant's happiness in terms of the selected place, while the second is related to the level of tolerance or rejection of refugees by the hosting community. Even though there are parts of the society that will always reject refugees [37, 38], Philips [37] argues that integration is successful when the refugees are given access to good quality accommodation available in their place of arrival. Strang and Ager [35] also suggest refugees' integration does not necessarily depend on relocating to an area with people from the same ethnic group. The three discussed aspects can be integrated into a model to mimic spatial decision making and movement of individuals forced to migrate due to conflicts such as the ongoing in Syria.

3 Data and Methods

3.1 Data

Data used in this model were acquired from various sources (Table 1). Monthly death toll data with location information was acquired via the Syrian Observatory for Human Rights [39], one of the most used data sources for the war in Syria. The extent of the datasets encompass the area of the refugee camps (Fig. 2a). From the road dataset only the primary roads and the highways were kept. From the 2004 Syrian census only those cities with more than 3000 people were considered. The model also includes indicators about destination countries [40, 41].

Table 1 Data classes and sources

Data	Source	Format
Population	Syrian census 2004, UN website, CIA world factbook	Excel
Roads, railways, cities	Open street map	Shp (point)
Airports	Open flight	Shp (point)
Political map	Thematic mapping	Shp (polygon)
Elevation	USGS global multi-resolution terrain elevation data 2010 (GMTED 2010)	Raster
Ethnic groups	GREG	Shp (polygon)
Death toll	Syrian observatory for human rights	Shp (point)
Recipient countries information	World economic forum, reporters without borders	Excel

3.2 The Model

We developed an ABM to simulate the migration of Syrian refugees, using death toll as an indicator of the severity of the conflict at each location and time. The model is comprised of human population agents who actively adapt and compare their tolerance with perceived severity of conflict and decide whether to leave their homes or not. Once they decide to depart, the agents consider a variety of conditions including their wealth, and choose a destination as well as a means of transportation. Upon arrival at destinations, the agents consider factors such as the capacity and existing population in the case of a refugee camp and decide whether to seek refuge therein. The model is iterated with 1 month time steps beginning in March 2011 and ending in December 2015.

The hybrid model allows cells to store data that is used to create, modify or influence agents [42, 43]. Other studies [16] used a ratio of agents to run the model. In our model Syrian cities create population agents at a ratio of 1 agent per 1000 people. If they decide to leave their city, population agents become migrant agents, and on arrival and settlement at their destination, the model transforms them to refugee agents. Destination options (countries, cities, and refugee camps) are given a series of indicators of attractiveness. These include: ethics, press freedom, organised crime, security, life expectancy, higher education, quality of education, quality of infrastructure, trustworthiness, public institutions and efficiency of government [49, 50]. The model also includes Syrian cities, other country cities, countries, refugee camps, airports, roads, and railways. The model implementation was made using Netlogo [44]; Fig. 1 depicts the flowchart.

3.2.1 Conflict Zones

To avoid problems (notably, lack of reliable data) of modeling the entire Syrian conflict we defined conflict zones in our model as monthly-updated, 20 km-wide areas around cities where deaths have been registered by the Syrian Observatory for Human Rights (SOHR). The higher the death toll at a zone, the more dangerous it is perceived by population agents. The deadliest recorded attack in SOHR data is assigned a danger score of 100 as reference, and danger scores of other records are calculated based on the proportion of their respective death toll to the reference. At each time step, conflict score of each conflict zone is calculated using the following formula:

$$C(t) = \frac{1}{2}[C(t-1) + D(t)]$$

$$C(t) = D(t) \quad \text{for } t = 1$$

where *C(t)* is the perceived conflict score of a zone at month *t*, and *D(t)* is the danger score of that zone at month *t*.

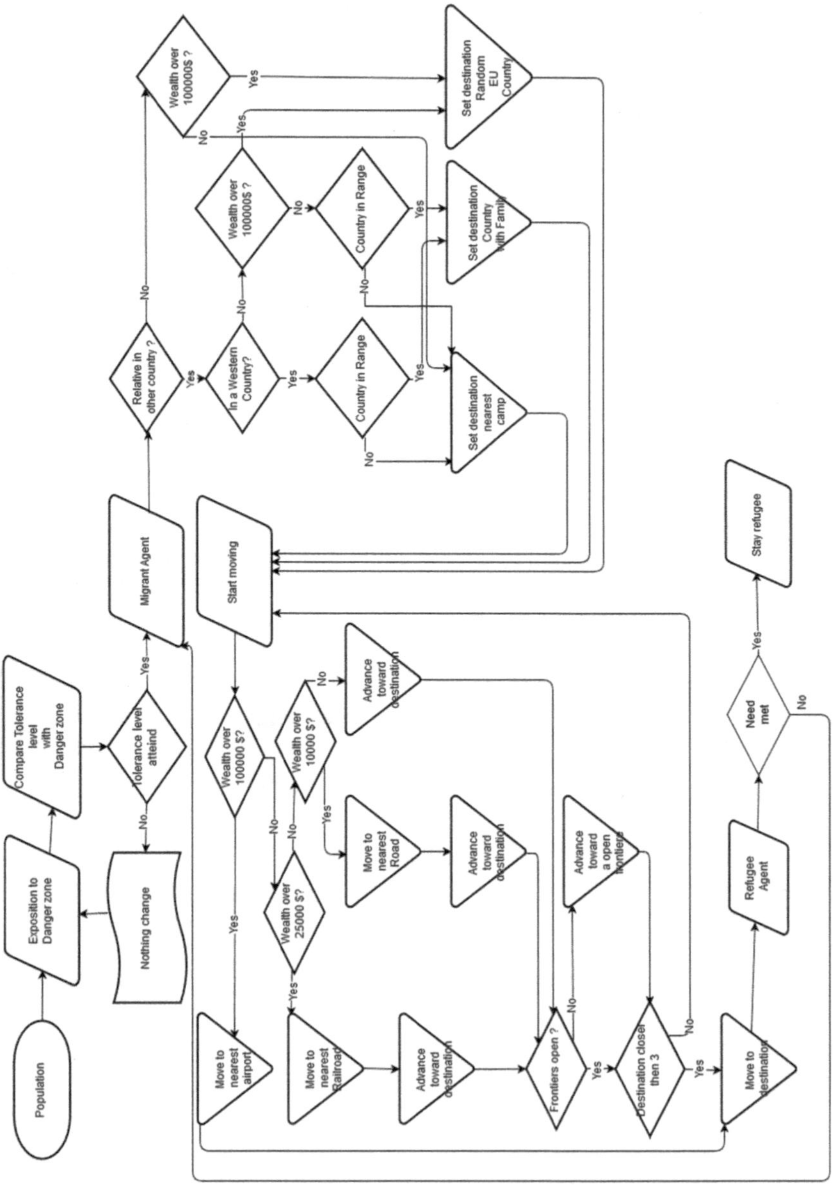

Fig. 1 Flowchart of refugee decision-making process

The inclusion of the perceived conflict score of the *previous* month (which in turn involves the effect of the month before, and so on) in the above formula enables the agents to have *memory*. The behaviour of the model will therefore be path-dependent. An implication of the above is that an agent may decide to leave even with a reduction in the danger score compared to previous time-step [45, 46].

3.2.2 Tolerance and the Decision to Leave

To assess populations' tolerance to conflict, a series of variables are associated to each agent [14]. Those include religion, ethnicity, wealth, age, sex, and familial status, the combination of which serves to determine the likelihood of an agent to leave [10, 47, 48]. In this study, due to unavailability of data, these parameters were randomly assigned. The model also considers a cumulative stress factor that reduces tolerance levels as the conflict continues. The decision rule for the first part of the model is based on a simple comparison: if the perceived severity of conflict exceeds tolerance, the population agent decides to leave, and becomes a migrant agent.

3.2.3 Destination Choice

Upon creation, migrant agents choose their destinations based on comparisons of their preferences with qualities of options available given their conditions. For example, a family member abroad could provide shelter and as such influences the choice of the migrant agent [36, 47, 49]. Moreover, wealthier migrants, can choose better means of transportation [49] and consider longer ranges. If no destinations exists for the expected criteria, the migrant agent chooses the nearest refugee camp.

3.2.4 Migration

Migrant agents choose a means of transportation depending on their wealth. On their journey, agents avoid cells with high danger scores and—especially for walking agents of very young and very old ages—cells with high slopes. Each migrant has a health score, which deteriorates with time and also in danger zones. Agents who lose all their health score, die. These features are to simulate the hardship of the migrants' journey [5, 6].

3.2.5 Arrival

Upon arrival at its destination, a migrant agent chooses whether to stay there and become a refugee agent. Overcrowding is a factor influencing such choice. Other important factors include infrastructure, sanitation and security [50, 51].

3.3 Model Parametrization

A challenge in the development of the model was the unavailability of data to set values of numerous influential parameters. We addressed this challenge by using information from relevant literature, particularly to identify parameters of highest importance, extract ranges of variation, and obtain an ordinal basis for categorizing and prioritizing levels of parameters for which a value is not available.

We acknowledge that the decision to leave is not a sudden event but the result of accumulation of pertinent factors [30, 32]. In this case, religion and ethnicity parameters were adjusted so as to make agents more tolerant when they are in regions of their own ethnicity [8]. Moreover, literature highlights that older people have more difficulty in the journey, and that they are more likely to stay in their home town longer [47, 52]. Accordingly, values of age parameters in the model were adjusted such that agents of higher age, choose departure later than others. Also, males generally have a better experience during the journey [53], therefore, the gender weight parameter is adjusted to generate more male migrants. Finally, literature notes that a single agent will be better suited to move as it is easier to travel alone than with a group [54], and that has been used as the basis for adjustment of age parameter of the model.

4 Results

4.1 Model Output

Figure 2b shows snapshots of migrant agent concentration at five temporal points. It is noticeable that the concentration of refugees in Jordan is greater than the in Turkey. Moreover, populations of Homs and Hama have fled the most. Refugee agents prefer camps far from danger zones. If we look at the Latakia region which is a regime stronghold we see a surge in migrants in that region between March and December 2015 which corresponds to the time where Russia first intervened [55].

Figure 3 shows migration pathways. We can see that a large number of migrants have fled the regions of Homs and Damascus. However this is not the case for the region of Aleppo. This can be explained by the danger level. Since Aleppo was near constant state of siege from the beginning of the conflict the danger level across this area is always high, which can refrain the migrant agents from moving. This represents the reality of many Syrians being trapped in their city or village, unable to flee due to the surrounding battle.

We can note many movements from Syria to Iraq. These could correspond to Kurd migrants and refugees fleeing northern Syria to safer zones in Iraq's Kurdistan. It is also observed that a lot of agents choose to move toward Jordan. We can also see some migrant movement from the Der-el-Zor region towards the Ambar Camp.

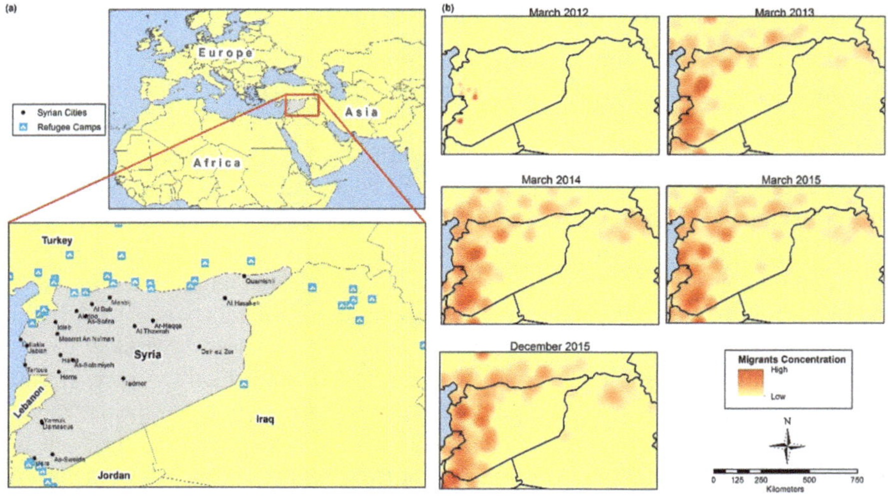

Fig. 2 (**a**) Study area and refugee camps; (**b**) Simulated migration heat map

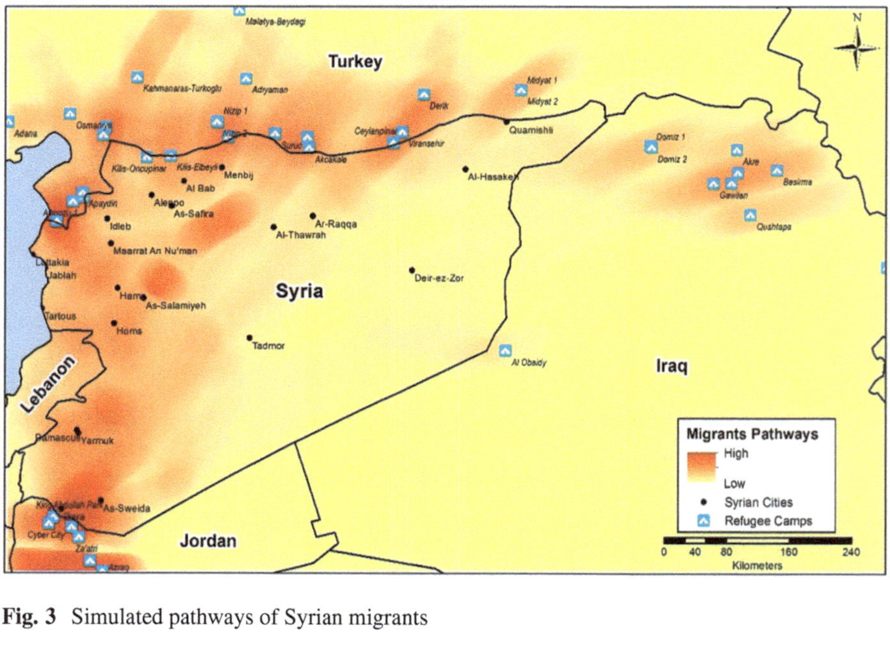

Fig. 3 Simulated pathways of Syrian migrants

4.2 Model Testing and Validation

Our model uses data on location, month, and number of deaths to simulate the consequent refugee flows. Due to unavailability of much of other data, the model cannot exactly replicate the real world. However, assuming that the death toll is a

Table 2 Quantity change and allocation tests

	Quantity	Allocation				
	Proportion change-2015	Iraq	Jordan	Lebanon	Turkey	Sum
UNHCR data	0.2661	0.0099	0.0442	0.0514	0.8946	1
Model output	0.3731	0.0927	0.2244	0.0003	0.6826	1
Minimum allocation agreement		0.0099	0.0442	0.0003	0.6826	0.7370

pertinent variable contributing to the situation, we expect our model to be able to show effects and dynamics comparable to those of the real world in terms of flows of refugees in response to the same spatiotemporal changes in the pertinent variable.

The UNHCR provides data on the number of refugees in its camps and neighboring countries [51]. We use these data to test the model, by comparing model outputs and real data in terms of quantity and allocation of change [56] between two time frames—December 2014 and December 2015. The variables that we used for model validation and testing are the rate of change in refugees arriving at neighboring countries (for quantity accuracy measurement), and the division proportions of new refugees among neighboring countries (for allocation accuracy measurement). In other words, we argue that the model generates new refugees each year and distributes them in the neighboring countries, and we perform tests to compare such generation rate and distribution shares with reference data from UNHCR [57, 58]. Table 2 shows the results of the tests.

The model shows a higher proportion of new refugees in 2015 compared to UNHCR data. This may be due in part to lack of detailed model design data, and to some extent to underestimation of the number of refugees in UNHCR records. As for allocations, the model shows lower proportions in Iraq and Lebanon, and higher in Jordan. These differences may be due to lack of data on spatial distribution of ethnicities. We used the sum of minimum agreements as a measure of allocation accuracy, ranging from 0 to 1. It must be noted that the above are results of 46th to 57th iterations (months) of the model, with death toll as the only input being updated. Moreover, while the model simulates one cause for migration of refugees, the observed data—which is the reference for testing the model—is aggregated and includes consequences of other possible causes as well.

5 Conclusion

We developed an agent-based model of violence induced migration. By using a simple variable—death toll—to simulate conflict zones, we built a model to generate migration patterns of refugees despite lack of reliable information on the details of the political and humanitarian conflict. Visual inspection of model outputs and comparison with observed data enabled us to better understand the model's capabilities and limits.

The model presented in this chapter is based on the idea that occurrence of violence at a location can cause refugee migration. Census data was used as input for spatial distribution of population. A highlight of the model design is the definition of conflict zones around locations of violence. Model testing was based on aggregate reported sums of refugees in neighbouring countries. The above aspects could also be considered for improvement and extension of future work. Regarding theory, violence-induced migration literature could suggest additional causes and mechanisms. As an example of improvement of inputs, socio-economic data with spatial distribution could, if available, replace some model assumptions. To improve the design, new models could be developed with different sizes of conflict zones, and compared to find the most realistic and reasonable amongst them. The outputs of our model could serve as information tool for humanitarian agencies in order to quickly prepare to receive refugees in cases of forced migrations, specifically in the case of the Syrian civil war. Future efforts could be made into finding reference data and developing tests that are more closely related to the model and that will help in its validation.

Acknowledgments Thanks to Dr. Anthony Jjumba for his valuable feedback and editing of this chapter.

References

1. Heather PJ (2016) Refugees and the Roman empire. J Refug Stud 30(2):220–242
2. Carling J (2015) Humanitarian crises and migration: causes, consequences and responses. J Refug Stud 28:138–140
3. Fargues P, Fandrich C (2012) Migration after the Arab Spring. MPC Research Report
4. Fargues P (2014) Europe must take on its share of the Syrian refugee burden, but how? Policy briefs, pp. 1–5
5. Fargues P, Fandrich C (2012) The European response to the Syrian refugee crisis–What next? European University Institute, p. 35
6. Kinninmont J (2014) The Syria conflict and the geopolitics of the region. IEMed Mediterranean Yearbook
7. Zetter R (2007) More labels, fewer refugees: remaking the refugee label in an era of globalization. J Refug Stud 20:172–192
8. Schmeild S (1997) Exploring the causes of forced migration: a pooled time-series analysis, 1971–1990. Soc Sci Q 78(2):284–308
9. Milner J (2014) Introduction: understanding global refugee policy. J Refug Stud 27:477–494
10. Oliver-Smith A (2009) Nature, society, and population displacement: Toward an understanding of environmental migration and social vulnerability. InterSecTions No. 8. Institute for Environment and Human Security (UNU-EHS), Bonn.
11. Sokolowski JA, Banks CM (2014) A methodology for environment and agent development to model population displacement. In: Proceedings of the 2014 symposium on agent directed simulation. pp. 3:1–3:11
12. Jackson IC (1991) The 1951 convention relating to the status of refugees: a universal basis for protection. Int J Refug Law 3:403–413
13. Makowsky M, Tavares J, Makany T, Meier P (2006) An agent-based model of crisis-driven migration. In: Proceedings of the Complex Systems Summer School, New Mexico

14. Smith C, Wood S, Kniveton D (2010, December) Agent based modelling of migration decision-making. In Proceedings of the European workshop on multi-agent systems (EUMAS-2010)
15. Kniveton D, Smith C, Wood S (2011) Agent-based model simulations of future changes in migration flows for Burkina Faso. Glob Environ Chang 21:S34–S40
16. Hassani-Mahmooei B, Parris BW (2012) Climate change and internal migration patterns in Bangladesh: an agent-based model. Environ Dev Econ 17:763–780
17. Rahman ABMZ (2009) Climate change, migration and conflict in Bangladesh: a view from the ground. IOP Conf Ser Earth Environ Sci 6:562003
18. Tang W (2008) Simulating complex adaptive geographic systems: a geographically aware intelligent agent approach. Cartogr Geogr Inf Sci 35:239–263
19. Manson SM (2001) Simplifying complexity: a review of complexity theory. Geoforum 32:405–414
20. Haken H, Mikhailov A (eds) (2012) Interdisciplinary approaches to nonlinear complex systems. Spinger-Verlag, New York, NY
21. Mainzer K (2007) Thinking in complexity: the computational dynamics of matter, mind, and mankind. Springer, New York, NY
22. Sprock T, McGinnis LF (2014) Modeling population displacement in the Syrian city of Aleppo. In: Proceedings of the 2014 Winter Simulation Conference, pp. 252–263
23. Collins AJ, Frydenlund E (2016) Agent-based modeling and strategic group formation: a refugee case study. In: Proceedings of 2016 Winter Simulation Conference, pp. 1289–1300
24. Epstein JM (2007) Generative social science: studies in agent-based computational modeling. Princeton University Press, New Jersey
25. Pellegrini PA, Fotheringham AS (2002) Modelling spatial choice: a review and synthesis in a migration context. Prog Hum Geogr 26:487–510
26. Henry S, Boyle P, Lambin EF (2003) Modelling inter-provincial migration in Burkina Faso, West Africa: the role of socio-demographic and environmental factors. Appl Geogr 23: 115–136
27. Yıldırım S, Yurtdaş GT (2016) Social construction of Syrian refugees in daily speech in Turkey: interpretative repertoires and social media. Middle East Journal Refugee Studies 1: 103–122
28. Taylor JE, Filipski MJ, Alloush M, Gupta A, Irvin R, Valdes R (2016) Economic impact of refugees. PNAS 113(27):1–5
29. Elizabeth B, Dunn C (2016) Refugee protection and resettlement problems. Science 352(6287):772–773
30. Oliver-Smith A (2012) Debating environmental migration: society, nature and population displacement in climate change. J Int Dev 24:1058–1070
31. Milner J (2014) Can global refugee policy leverage durable solutions? Lessons from Tanzania's naturalization of Burundian refugees. J Refug Stud 27:553–573
32. Klabunde A, Willekens F (2016) Decision-making in agent-based models of migration: state of the art and challenges. Eur J Popul 32:73–97
33. Ajzen I (1991) The theory of planned behavior. Organ Behav Hum Decis Process 50:179–211
34. Moore WH, Shellman SM (2006) Refugee or internally displaced person?: to where should one flee? Comp Polit Stud 39:599–622
35. Strang A, Ager A (2010) Refugee integration: emerging trends and remaining agendas. J Refug Stud 23:589–607
36. Charteland G (2008) A quest for family protection: the fragmented social organisation of transnational Iraqi migration. In: Academy B (ed) Displacement and dispossession: force migration in africa and the Middle East, p 16
37. Phillips D (2010) Minority ethnic segregation, integration and citizenship: a european perspective. J Ethn Migr Stud 36:209–225
38. Awad I (2014) Population movements in the aftermath of the Arab awakening: the Syrian refugee crisis between regional factors and state interest, pp 24–39
39. Syrian Observatory for Human Rights. http://www.syriahr.com/en/
40. Schwab K (2012) World E.F.: the global competitiveness report 2012–2013, Geneva

41. World press freedom index (2016) Reporters Without Borders. https://rsf.org/en/ranking/2016
42. Gulden T, Harrison J, Crooks A (2011) Modeling cities and displacement through an agent-based spatial interaction model. In: The Computational Social Science Society of America Conference
43. Chen SH, Jakeman AJ, Norton JP (2008) Artificial intelligence techniques: an introduction to their use for modelling environmental systems. Math Comput Simul 78:379–400
44. Wilensky U (1999) NetLogo. https://ccl.northwestern.edu/netlogo/
45. Brown DG, Page S, Riolo R, Zellner M, Rand W (2005) Path dependence and the validation of agent-based spatial models of land use. Int J Geogr Inf Sci 19:153–174
46. O'Sullivan D (2004) Complexity science and human geography. Trans Inst Br Geogr 29: 282–295
47. Stefanovic D, Loizides N, Parsons S (2014) Home is where the heart is? Forced migration and voluntary return in Turkey's Kurdish regions. J Refug Stud 28:feu029
48. Melander E, Öberg M (2007) The threat of violence and forced migration: geographical scope trumps intensity of fighting. Civ Wars 9:156–173
49. Adhikari P (2012) The plight of the forgotten ones: civil war and forced migration. Int Stud Q 56:590–606
50. Melander E, Öberg M (2006) Time to go? Duration dependence in forced migration. Int Interact 32:129–152
51. Willekens F (2013) Agent-based modeling of international migration. In: Research plan for independent research group. Max Planck Institute for Demographic Research, Rostock, pp 1–19
52. Webber SC, Porter MM, Menec VH (2010) Mobility in older adults: a comprehensive framework. Gerontologist 50:443–450
53. de Haas H, van Rooij A (2010) Migration as emancipation? The impact of internal and international migration on the position of women left behind in rural Morocco. Oxf Dev Stud 38:43–62
54. Willekens F, Massey D, Raymer J, Beauchemin C (2016) International migration under the microscope. Science 352(80):897–899
55. Quinn B (2016) Russia's military action in Syria–timeline. The Guardian. https://www.theguardian.com/world/2016/mar/14/russias-military-action-in-syria-timeline
56. Pontius RG, Millones M, Gilmore R (2011) Death to kappa: birth of quantity disagreement and allocation disagreement for accuracy assessment. Int J Remote Sens 32:4407–4429
57. United Nations High Commissioner for Refugees (2016) Global Trends: Forced displacement in 2015. http://www.unhcr.org/576408cd7.pdf
58. United Nations High Commissioner for Refugees (2017) UNHCR Syria regional refugee response. http://data.unhcr.org/syrianrefugees/regional.php

Automated Extraction of Movement Rationales for Building Agent-Based Models: Example of a Red Colobus Monkey Group

Raja Sengupta, Colin C. Chapman, Dipto Sarkar, and Sarah Bortolamiol

Abstract The study of animal movement has gained impetus in recent years with improvements in telemetric technologies which enable high resolution tracking, providing researchers with a wealth of animal "big-data". Coupling such movement data with information about the environments in which the animal moves provides a rich data source that can be exploited to understand an animal's rationale for movement, which in turn can be used to extract "rules" that govern movement. The extraction of rules can be done using spatial, statistical and machine learning techniques. Once the rules replicating patterns and predictors of movement have been "discovered", they can be subsequently used to build simulation models (ABMs) to mimic in-silico the behaviours of both individuals and groups of animals. We use field data collected by tracking Red Colobus (*Procolobus rufomitratus*) monkey groups from Kibale National Park, combined with land cover and terrain information, to show how this might be achieved.

Keywords Agent-based models • Machine learning • GIS • Movement ecology • Big data

---

R. Sengupta (✉)
Department of Geography, School of Environment, McGill University, Montreal, QC, Canada H3A 0G4
e-mail: raja.sengupta@mcgill.ca

C.C. Chapman
School of Environment, McGill University, Montreal, QC, Canada H3A 0G4

Department of Anthropology, McGill University, Montreal, QC, Canada H3A 0G4
e-mail: colin.chapman@mcgill.ca

D. Sarkar
Department of Geography, McGill University, Montreal, QC, Canada H3A 0G4
e-mail: dipto.sarkar@mail.mcgill.ca

S. Bortolamiol
Department of Geography, McGill University, Montreal, QC, Canada H3A 0G4

Department of Anthropology, McGill University, Montreal, QC, Canada H3A 0G4
e-mail: sarah.bortolamiol@mail.mcgill.ca

L. Perez et al. (eds.), *Agent-Based Models and Complexity Science in the Age of Geospatial Big Data*, Advances in Geographic Information Science,
DOI 10.1007/978-3-319-65993-0_5

1 Introduction

Agent-Based Models (ABMs) have been used extensively to explore the impact of animal movement patterns across space-time and predict environmental outcomes. As an example, ABM simulations of red colobus (*Procolobus rufomitratus*) monkey groups in Kibale National Park, Uganda, suggested that fragmentation of landscapes combined with animal movement strategies allow for the emergence of hotspots for zoonotic diseases [1]. However, the movement rationale expressed in such ABMs have thus far been based on expert knowledge about the behaviour of the Red Colobus monkeys, which were subsequently converted to rules.

With the advent of tracking technologies such as GPS tags, there has been a concomitant rapid rise of animal movement studies generating an enormous volume of valuable tracking data [2]—an example of "big data". This provides a significant opportunity to utilize this widespread availability of movement data and extract the rationale behind the movements, and to convert these into agent-rules. Here, we propose that the availability of such large datasets with high spatial and temporal granularity (both animal and human) can be combined with other GIS data and methods for automated extractions of movement rules. Tested rationales could be preferred habitats, avoidance of high risk predator or disease areas, territorial defense, and social behaviour. This augments the expert's interpretation, which was traditionally based on field observations. Additionally, success in identifying the rationales for movement can be used for parameterization and for calibration of ABM model output [3–5].

2 Extracting Movement Rationales from Data

Over two decades ago, Rodgers & Anson [6] had prophetically suggested that "GPS-based animal-location systems will set a new standard for habitat-resource utilization studies of large animals over the next five to 10 years". This availability of high resolution movement data, particularly those collected via GPS telemetry (i.e., sequence of GPS locations), has given rise to the field of "movement ecology" [7]. Additionally, the Max Planck Institute of Ornithology has developed a free online database, Movebank (movebank.org) that allows researchers interested in animal movement to "manage, share, protect, analyze, and archive their data". Current studies range from estimating the home ranges of animals to understanding the space use and detailed movements of animals [8]. Furthermore, new methods have been developed to analyze the data to understand the unknown rules followed by the study animals [9, 10]. The broad goal of movement ecology is to study the processes that cause and influence movement in animals [11]. These processes are diverse, with suggestions that individual mechanisms such as spatial memory, internal time measures, communication, and reliance on co-specifics are all factors that underlie movement behaviour [12–15]. Additionally, to coordinate the nature and timing of their activities (including movement), interactions amongst individuals is necessary for most animals living in groups [16]. Since social hierarchies and predation risk vary among species and individuals and resulting in individual-specific strategies, this further complicates our understanding of movement dynamics.

In its most elementary stage, Nathan et al. [7] suggest that the movement of an individual organism occurs due to the interplay of four mechanistic components: its internal state, its motion capacity, its navigation capacity and external factors. Internal states are difficult to capture solely from "big data" at present. However, we propose that attempts can be made to extract rules about motion, navigation and external (environmental) factors. Motion and navigation, for example, manifest themselves as the direction, magnitude and periodicity of movement, all of which can be extracted from time-series location information [11, 17–19]. Recently, it has been suggested that there are common movement strategies across taxa (although such generalizations can be quickly disputed) [20, 21], further bolstering the argument that motion by itself can be quantified and extracted as rules. Moreover, information about navigation can be gleaned by studying external factors (e.g., land use-land cover, topography) to identify environmental reasons that drive movements (e.g. navigation is controlled by availability of food sources but limited to specific areas due to slope).

New spatial methods have focused on analyzing the relative periodicity and directionality of movement as an important and integral part of a broader framework of movement-related studies in GIScience [11, 17, 18]. Specifically, pattern and cluster methods can help identify similarity of movement behavior or locate places of repeat interaction or use [22]. Understanding these "episodal movements" are critical to capture repeat patterns in the behaviour of a moving point object [17, 23]. When integrated with distance, it can also provide information about similarity of movement patterns for a pair of moving objects.

Additionally, several applications have been coupled with spatial analysis methods to provide a better understanding of animal behaviours from an ecological perspective [24–26]. Very useful software packages have been built to exploit information from telemetry-based movement data combined with spatial (GIS-based) datasets (e.g., datasets on percent canopy cover, elevation, water bodies etc.). For example, Geospatial Modelling Environment [27] analyses animal movements considering the surrounding ecosystem, and allows these movements to be decomposed into component localized movements that can be correlated with environmental or habitat information. Such specialized open source software augment the analysis provided by traditional GIS methods. Importantly, they allow researchers to understand the movements in the context in which they are occurring (e.g., fragmented landscapes with or without corridors). The Environmental Data Automated Track Annotation System (EnvDATA) within Movebank is one such software that allows environmental data obtained from remotely sensed satellite information to be attached onto Movebank's data locations [11, 28]. This then facilitates a greater understanding by allowing the movement to be contextualized with respect to the environment in which it occurred.

3 An Example of Red Colobus (*Procolobus rufomitratus*) Monkey Group Movements in Kibale National Park, Uganda

To simplify movement rationale, we seek to consider three main drivers of movement patterns: (1) availability of food resources, (2) social factors (e.g., territoriality, mating opportunities), and (3) predation risk. Additionally, several external and observable factors (e.g., percent canopy cover, elevation, water bodies) can be considered while determining movement behavior. This allows predicting the percentage of the variation in movement patterns explained by each driver. Such variation can then be used to derive rules that may be pertinent to deciphering movements in a variety of contexts.

In order to demonstrate how automated rule extraction for ABM development could proceed, we analyze a 1.5 year snapshot (30 March, 2011–15 Sept, 2012) of a red colobus monkey group movement in Kibale National Park (KNP), Uganda (Fig. 1). The dataset includes sighting location co-ordinates by continuously following one group of red colobus monkeys living inside KNP on a daily basis. The GPS points were collected every 15 min by a research assistant located amidst

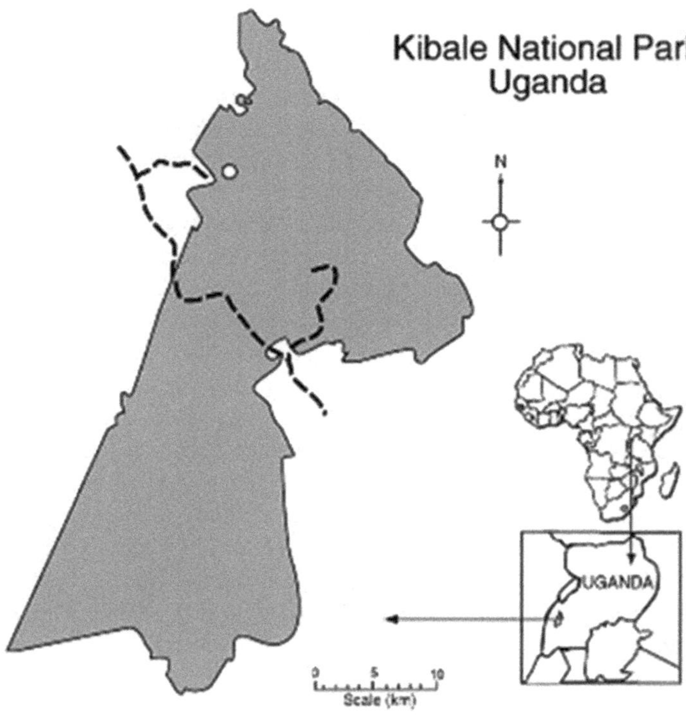

Fig. 1 Location of study area: Kibale National Park, Uganda

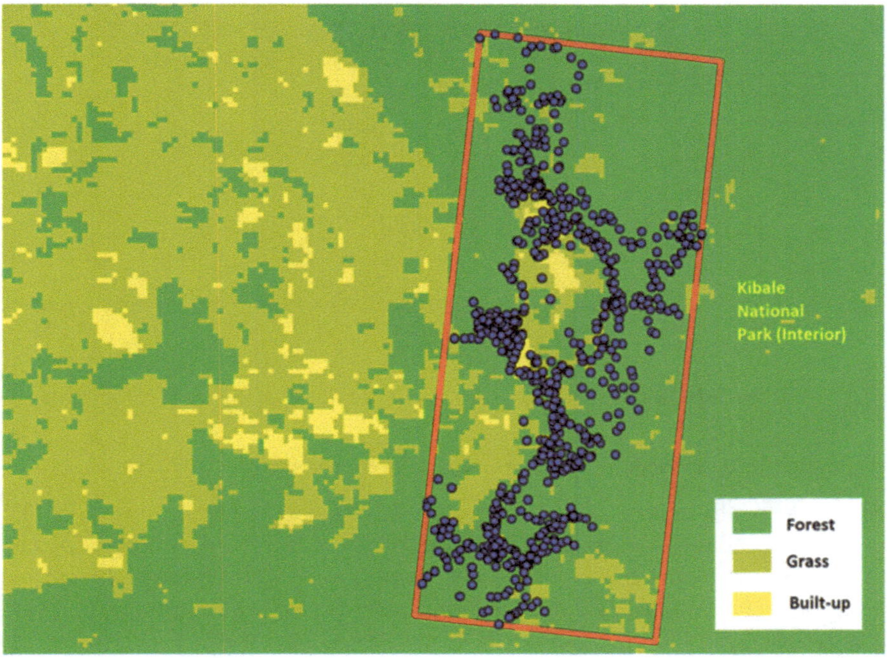

Fig. 2 Distribution of Red Colobus observations (girded by a Minimum Bounding Rectangle)

members of this group. A total of 743 observations were collected in this time period (Fig. 2), and all observations fall within an approximately 600 m × 1500 m bounding rectangle highlighted in red. Also red colobus are not territorial, they are relatively small size mammals and live in social groups that do not move in search of mates [29]. Further, this set of points were selected for analysis because of their relative continuity (i.e., lack of gaps in data collection), and because of the fact that no predation was observed during this period. This is important because predation can significantly alter movement characteristics in red colobus [30, 31]. Thus, all movement seen during this period is likely solely because of foraging strategies employed by the group.

To direct this work, movement rationales were broken down into two categories, one related to the movement itself, and the other to underlying environmental factors controlling movement. For the first category of "movement rules", characteristics of movement such as initiation, distance, and direction can be extracted from the analysis of big movement data and used to suggest an agent's probable motion. The second, "constraining rules" analyze if the new location proposed by "movement rules" is viable based on environmental factors. Hence, questions relating to the two categories can be specified as (Fig. 3):

Movement rules: How frequently does the group move? And once the group is in motion, how far and in which direction does it move?

Constraining Rules: What is the most common environmental factor (e.g., percent canopy cover, elevation, water bodies) that puts spatial bounds on the groups' observed location?

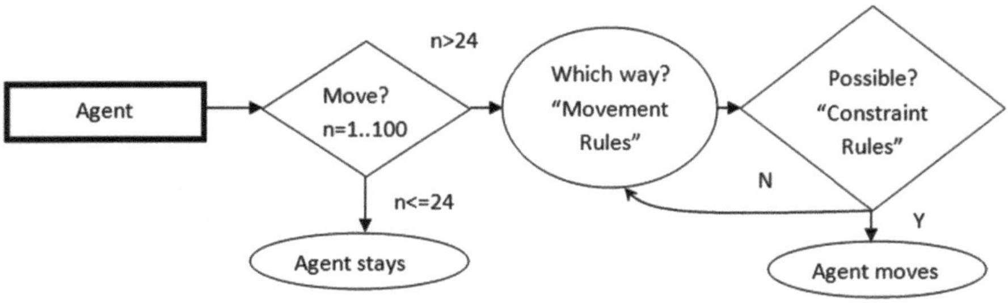

Fig. 3 Decision-making steps taking by an agent based on "movement" and "constraint" rules

3.1 Movement Rules

How frequently and which way (distance, direction) does the group move?

During the period of observation (taken every 15 minutes), the group moved 76.03% of the time, and was consequently stationary the rest (23.97%) of the time. Overall, the group's movement was normally distributed (Fig. 4) with a mean of 29.5 m and a standard deviation of 27.86 m. However, there were longer transects up to 303 m during the observation period. These infrequent yet important longer movements show up in a detailed time series analysis of movement data [32], particularly via a spectral analysis where the periodogram shows a prominent seasonal trends every ten readings (Fig. 5). There was no observed correlation between frequency and distance of movement.

To convert these analyses into "movement rules" governing motion of agents (where the agent is the group), three components of any movement can be considered: initiation, distance and direction. Initiation, which is the start of a movement following a sedentary period, should be proportional to the time where movement was observed (76.03%). For this study, this can be controlled via a rule that depends on a random function, e.g., pick an integer between 0–100, and initiate movement if the random number exceeds 24 (Fig. 3). An additional rule can then randomly select the distance to move as a function of the mean and standard deviation of observed data (i.e., the normally distributed observations as seen in Fig. 4). It should also allow for longer transects to be included cyclically every ten time steps (and with some randomness of ± 1–2 time steps, as evidenced by the spectral analysis in Fig. 5). An analysis of the direction of movement did not yield any prominent trend for this dataset, i.e., there seems to be no specific preferences (Fig. 4). The direction of movement can therefore be randomly selected to be from 0 to 359°. Together, these rules specify the initiation, distance and direction of movement as a function of observed parameters.

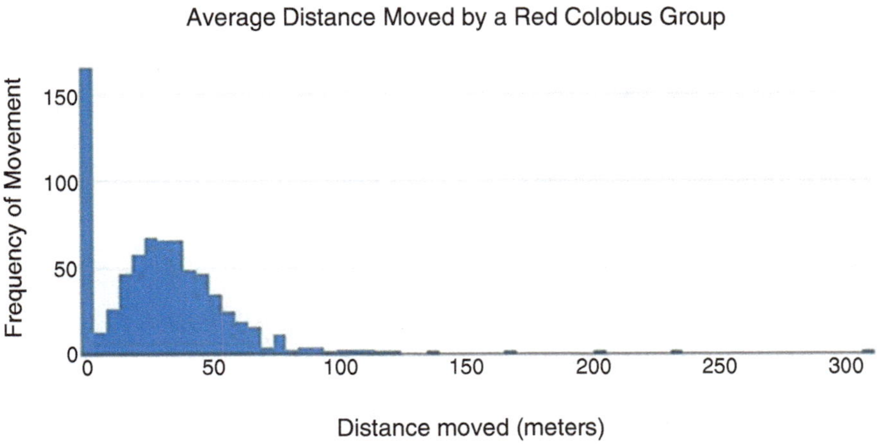

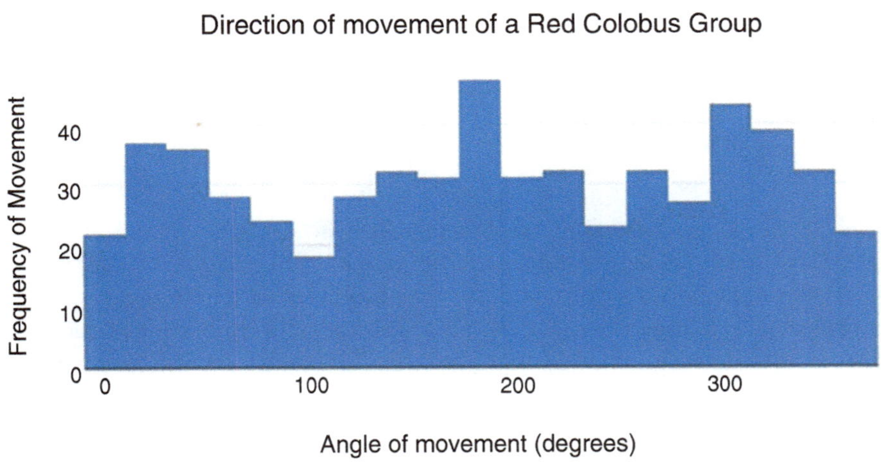

Fig. 4 Distribution of movement frequency of Red Colobus group

3.2 *Constraining Rules*

> What are the most common environmental factors (e.g., percent canopy cover, elevation, water bodies) that puts bounds on the groups' motion?

The "movement rules" derived by analyzing the data on when, how far, and in what direction the animals moved can readily be used to inform agents in an ABM. However, there is no check to see whether the move itself would be possible in a real-world setting. For example, movement rules may suggest a location far away from a forest edge as they do not consider land cover, but in the real-world the group may never move there due to safety concerns and other factors. Environmental factors such as availability of food and water resources, and other constraints

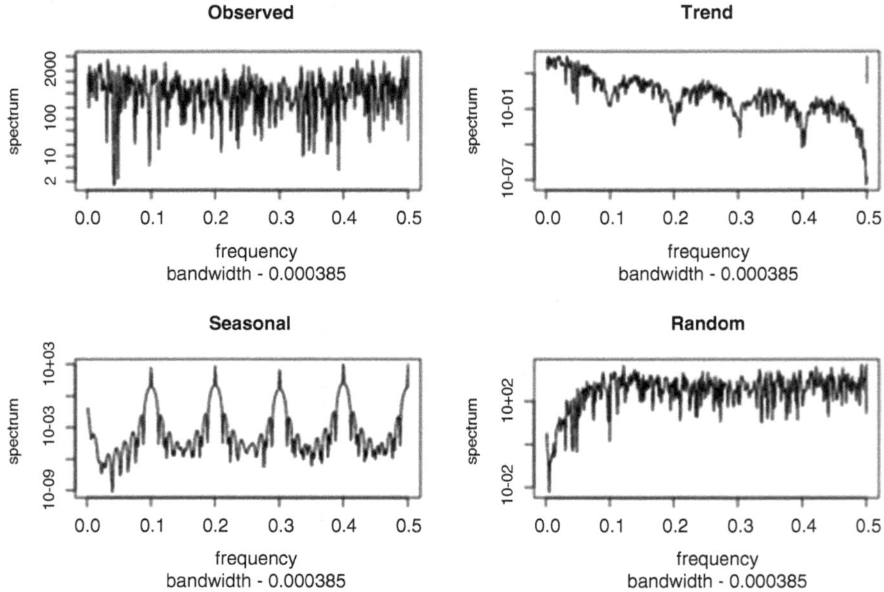

Fig. 5 Seasonality detected in the periodogram indicating pattern in movement trends [32]

such as elevation, often limit the exact movement strategies of most species, including the red colobus. We therefore propose a set of additional "constraining rules" that evaluate the appropriateness of the new location for the agent based on environmental characteristics (Fig. 3). These rules act as a check on the directional rules generated above, i.e., if the suggested "new" location for an agent group does not meet the criteria generated from analysis of environmental characteristics, then the agent is not allowed to move, but a new movement rule is generated. The agent is only allowed to move IF the new location proposed by movement rule meets the criteria of environmental factors, as specified below.

To evaluate the controls exerted by environmental factors, we selected a group of GIS-based layers guided by expert knowledge about common ecological constraints [33, 34] on the red colobus groups. These were analyzed using ArcGIS 10's Spatial Analyst functions [35]. Specifically, we utilized a land cover (LC) map (obtained from supervised classification of SPOT imagery; classified as Forest, Grass/Swamp and Built up areas) as well as a Digital Elevation Model (DEM of 90 m resolution, obtained from the SRTM Shuttle Radar mission) of the study area. Following analyses were conducted using the base information: (1) extract a raster layer indicating distance (in meters) away from open areas (i.e., Grass/Swamp and Built-up; as specified in the reclassified SPOT image); (2) derive slope in degrees from DEM using the "Slope" function; and (3) associate the values from these four raster layers (LC, DEM, distance raster, and slope) with the 743 point observation locations. The last stepstores the extracted values of the four raster layers in four corresponding attribute fields for each of the 743 observation locations: LULCVAL with values 0-forest 1-grass/swamps, 2-built-up areas; DEMVAL; DISTVAL; and SLOPEVAL.

To test if there are indeed environmental controls on movements, a data mining software Weka 3.8 [36], and its M5 pruned model tree with default values, was utilized. The M5 is a decision tree classifier with linear regression functions at the leaves [37, 38]. If a presence/absence dataset is provided to it, it can generate a decision tree that "classifies" the presence/absence (dependent variable) as a function of the independent variables—which in turn are selected using linear regression at the leaf level. To run the classifier, an additional 743 random points were generated as "absence points" (using a "create random points" function) that serve as the null hypothesis. As with the 743 observation locations (now denoting "presence"), values from the four environmental layers (LC, DEM, distance and slope) were also associated with these newly generated random points. The dependent variable is now denoted by a 0 for random/absence and 1 for presence.

Initially, the "absence" points were randomly generated within the red bounding box, as this was assumed to be the region of occurrence for the Red Colobus group (Fig. 2). The resulting decision tree (shown in Fig. 6a) suggests that a distance of less than 58.125 meters from open areas (grass/swamps and built up; DISTVAL <58.125) is the only deciding factor in the location of the monkey group. The "constraining rule" therefore is that an agent (representing the Red Colobus group) is allowed to move to a new location only within 58.125 m of open areas. Else another movement rule has to be fired, specifying a new distance and direction of movement. However, this single constraint of within 58.125 m produced by Weka 3.8 may be a result of constraining the random points to within the bounding box, where most of the land cover is forest (87.3% of observed locations, and 83.9% of the randomly generated points fell on forested areas). Given the fact that the Red Colobus group is not really territorial [29], a larger area was subsequently considered.

The consideration of a larger area (Figs. 6b and 7) suggested a more complex picture, with a distance of <100.12 m from open areas being considered the threshold for locations visited by the monkey group (the randomly generated points, on the other hand, were located at distances >100.12 m). Additionally, the actual Red Colobus observations were located inside forested areas (an LULCVAL of 0 indicates forests; with the decision tree suggesting that LULCVAL <0.5 indicates observed Red Colobus locations). This analysis indicates that any movement by the Red Colobus group must occur in forested areas that are within 100.12 m of grass and built-up areas, which are basically the forest edges (Fig. 7). The related constraining rules are therefore: allow the move (as specified by the highest value of 0.6 in Class 5; Fig. 6b) if the new location is less than 100.12 m from forest edge (DISTVAL <100.12) and is located on a land cover value of 0 (LULCVAL <0.5).

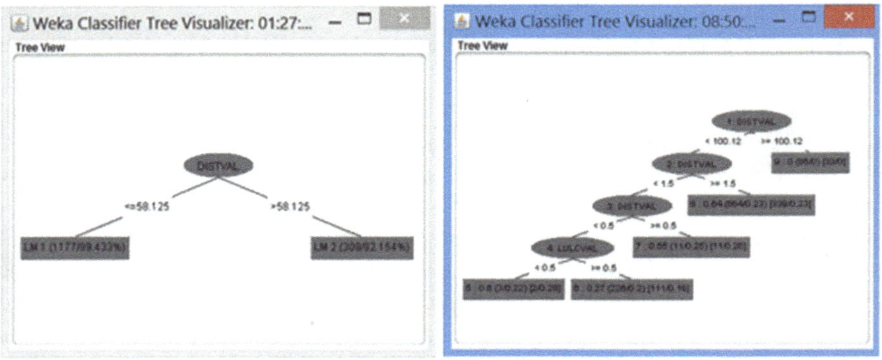

Fig. 6 (**a, b**) If the random points generated are restricted to the bounding rectangle in Fig. 2, distance from open areas (DISTVAL) <58.125 is the only variable controlling the location of observations. Comparing to a larger set of random points, the observed Red Colobus locations are <100.12 m from open areas, and always located in forested areas (LULCVAL of 0; LULCVAL <0.5)

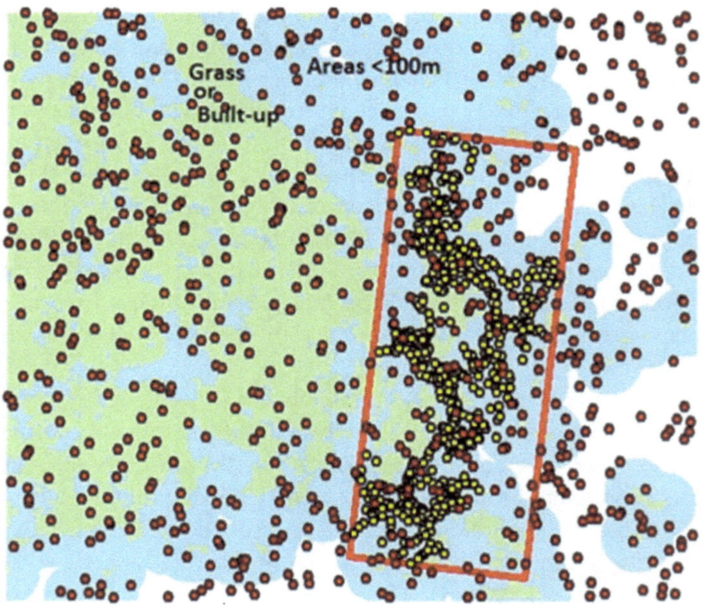

Fig. 7 The observed Red Colobus (*Procolobus rufomitratus*) (*yellow dots*) and random (*red dots*) locations superimposed on grass/built-up areas (*light green*), and areas <100 m (*light blue*) from them

4 Future Possibilities

As the availability of "big data" collected at a high spatial and temporal resolution grows, it opens up options for its analysis with spatial, statistical, and data mining and learning techniques to develop and refine the rules governing movement in ABMs. Movebank, for example, included data from 2484 studies across 548 taxa, and from 303 million locations. This represents an enormous wealth of data on animal movement patterns across species, spatial and temporal scales, and landscapes. It also opens the door for detailed analysis of the patterns and rationale for movement across numerous species, functional groups, habitat, landscapes, disturbance regimes, etc. Specifically, it becomes possible to derive rules that control the initiation, motion and navigation of individual agents, as well as to place constraints on the plausibility of certain movements. The attempts herein to develop standard methodologies using statistics and machine learning to extract rules from observational data meshes well with concurrent work elsewhere to automatically extract movement rules, as well as calibrate motion in ABMs [3–5]. In the future, such automated extractions are expected to replace or augment the heuristic knowledge of experts regarding animal movement, which had hitherto been the standard method for deriving rules [1, 26, 39, 40]. As a consequence, the future of ABM model development and calibration may increasingly depend on the extracting of meaningful patterns from a significant source of movement data. We provide one way forward towards achieving this objective by developing automated methods of extracting movement behaviors as "movement" and "constraining" rules, and representing the agents movement as an interplay of these two sets of rules.

References

1. Bonnell TR, Sengupta R, Chapman C, Goldberg T (2010) An agent-based model of red colobus resources and disease dynamics implicates key resource sites as hotspots of disease dynamics. Ecol Model 221:2491–2500
2. Kranstauber B, Cameron A, Weinzerl R, Fountain T, Tilak S, Wikelski M, Kays R (2011) The movebank data model for animal tracking. Environ Model Softw 26:834–835
3. Malleson N, Birkin M (2013) Estimating individual behaviour from massive social data for an urban agent-based model. In: Koch A, Mandl P (eds) Modelling social phenomenon in spatial context: geosimulation, vol 2. LIT Verlag, Munster
4. Torrens P, Li X, Griffin W (2011) Building agent-based walking models by machine-learning on diverse databases of space-time trajectory samples. Trans GIS 15(s1):67–94
5. Ward J, Evans A, Malleson N (2016) Dynamic calibration of agent-based models using data assimilation. R Soc Open Sci 3:150703
6. Rodgers A, Anson P (1994) Animal-borne GPS: tracking the habitat; GPS World, pp 20–32
7. Nathan R, Getz WM, Revilla E, Holyoak M, Kadmon R, Saltz D, Smouse PE (2008) A movement ecology paradigm for unifying organismal movement research. Proc Natl Acad Sci 105(49):19052–19059

8. Burdett CL, Moen RA, Niemi G, Mech LD (2007) Defining space use and movements of Canada lynx with global positioning system telemetry. J Mammal 88(2):457–467
9. Merrill S, Mech L (2003) The usefulness of GPS telemetry to study wolf circadian and social activity. Wildl Soc Bull 31(4):974–960
10. Ropert-Coudert Y, Wilson R (2005) Trends and perspectives in animal-attached remote sensing. Front Ecol Environ 3(8):437–444
11. Demšar U, Buchin K, Cagnacci F, Safi K, Speckmann B, Van de Weghe N, Weiskopf D, Weibel R (2015) Analysis and visualisation of movement: an interdisciplinary review. Move Ecol 3(1):1–24
12. Heinrich B (1988) Winter foraging at carcasses by three sympatric corvids, with emphasis on recruitment by the raven, *Corvus corax*. Behav Ecol Sociobiol 23:141–156
13. Boinski S, Garber PA (2000) On the move: how and why animals travel in groups. University of Chicago Press, Chicago
14. Janmaat KRL, Byrne RW, Zuberbuhler K (2006) Evidence for a spatial memory of fruiting states of rainforest trees in wild mangabeys. Anim Behav 72:797–807
15. Janmaat KRL, Chapman CA, Meijer R, Zuberbuhler K (2012) The use of fruiting synchrony by foraging mangabey monkeys: a 'simple tool' to find fruit. Anim Cogn 15:83–96
16. King AJ, Sueur C (2011) Where next? Group coordination and decision making by primate. Int J Primatol 32:1245–1267. doi:10.1007/s10764-011-9526-7
17. Sarkar D, Chapman C, Griffin L, Sengupta R (2015) Analyzing animal movement characteristics from location data. Trans GIS 19(4):516–534
18. Laube P, Wolle T, Gudmundsson J (2007) Movement patterns in spatio-temporal data. Encyclopedia of GIS
19. Long JA, Nelson TA (2013a) A review of quantitative methods for movement data. Int J Geograph Sci 27(2):292–318
20. Abrahms B, Seidel D, Dougherty E, Hazen E, Bogard S, Wilson A, McNutt J, Costa D, Blake S, Brashares J, Getz W (2017) Suite of simple metrics reveals common movement syndromes across vertebrate taxa. Move Ecol 5:12–23
21. Gao P, Kupfer JA, Zhu X, Guo D (2016) Quantifying animal trajectories using spatial aggregation and sequence analysis – a case study of differentiating trajectories of multiple species. Geogr Anal 48(3):275–291
22. Long J, Nelson T, Wulder M (2010) Regionalization of landscape pattern indices using multivariate cluster analysis. Environ Manag 46(1):134–142
23. Long JA, Nelson TA (2013b) Measuring dynamic interaction in movement data. Trans GIS 17(1):62–77
24. Patterson TA, Basson M, Bravington MV, Gunn JS (2009) Classifying movement behaviour in relation to environmental conditions using hidden Markov models. J Anim Ecol 78(6):1113–1123
25. Dodge S, Weibel R, Ahearn SC, Buchin M, Miller JA (2016) Analysis of movement data. Int J Geogr Inf Sci 30(5):825–834
26. Bonnell TR, Campenni M, Chapman C, Gogarten J, Reyna-Hurtado R, Teichroeb J, Wasserman M, Sengupta R (2013) Emergent group level navigation: an agent-based evaluation of movement patterns in a folivorous primate. PLoS One 8(10):e78264. doi:10.1371/journal.pone.0078264
27. Beyer HL (2012) Geospatial modelling environment (Version 0.7.2.0)
28. Dodge S, Bohrer G, Weinzierl R, Davidson SC, Kays R, Douglas D, Cruz S, Han J, Brandes D, Wikelski M (2013) The environmental-data automated track annotation (Env-DATA) system: linking animal tracks with environmental data. Move Ecol 1(1):1
29. Chapman CA, Chapman LJ, Gillespie TR (2002) Scale issues in the study of primate foraging: red colobus of Kibale National Park. Am J Phys Anthropol 117:349–363
30. Gebo DL, Chapman CA, Chapman LJ, Lambert J (1994) Locomotor response to predator threat in red colobus monkeys. Primates 35(2):219–223

31. Boinski S, Treves A, Chapman CA (2000) A critical evaluation of the influence of predators on primates: effects on group travel. In: Boinski S, Garber PA (eds) On the move: how and why animals travel in groups. University of Chicago Press, Chicago, pp 43–72
32. Wessa P (2013) Classical decomposition (v1.0.4) in free statistics software (v1.1.23-r7), Office for research development and education, URL http://www.wessa.net/rwasp_decompose.wasp/
33. Chapman CA, Wrangham R, Chapman LJ (1995) Ecological constraints on group size: an analysis of spider monkey and chimpanzee subgroups. Behav Ecol Sociobiol 36:59–70
34. Struhsaker TT (1975) The red colobus monkey. Chicago University Press, Chicago
35. ESRI Inc (2017) ArcGIS 10.3. http://www.esri.com/arcgis
36. University of Waikato (2017) WEKA 3.8, https://cs.waikato.ac.nz/weka
37. Quinlan JR (1992) Learning with continuous classes. In: 5th Australian joint conference on artificial intelligence, vol 92, pp 343–348
38. Frank E, Wang Y, Inglis S, Holmes G, Witten I (1998) Using model trees for classification. Mach Learn 32(1):63–76
39. Bonnell T, Chapman C, Sengupta R (2016) Interaction between scale and scheduling choices in simulations of spatial agents. Int J Geogr Inf Sci 30(10):2075–2088
40. Sengupta R, Sieber R (2007) Geospatial agents, agents everywhere .... Trans GIS 11(4):483–506

Wealthy Hubs and Poor Chains: Constellations in the U.S. Urban Migration System

Xi Liu, Ransom Hollister, and Clio Andris

Abstract Flows of people connect cities into complex systems. Urban systems research focuses primarily on creating economic models that explain movement between cities (whether people, telecommunications, goods or money), and more recently, finding strongly and weakly-connected regions. However, geometrically graphing the dependency between cities within a large network may reveal the roles of small and peripheral city agents in the system to show which cities switch regions from year to year, which medium-sized cities serve as collectors for large cities, and how the network is configured when connected by wealthy or deprived agents.

We propose a network configuration method called 'best friend' networks, where a node attaches to one preferential node, so that edges = nodes = n. Our case study is 20 years of migrants, sourced from the U.S. Internal Revenue Service, traveling between U.S. cities. In our networks, an edge is created to link a city to its most popular migrant destination city for a given year. The resulting configurations reveal closely connected "constellations" of cities comprised of chains, trees, and hub-spoke structures that show how urban regions are configured. We also show routing behavior within these networks to reveal that high-income migrants tend to flock to hub cities, while low-income migrants form local city chains via nearby movements.

Keywords Migration • Urban hierarchy • Economic systems • Regional science • Spatial interaction • Complex systems

1 Introduction

In an urban hierarchy, larger cities are connected to smaller cities with medium size cities as intermediaries. Within this network, goods, information, capital, flights, migrants, commuters, etc. flow through planar and non-planar veins, providing cities with valuable resources. While cities are often studied in terms of demographics and

---

X. Liu • R. Hollister • C. Andris (✉)
Department of Geography, The Pennsylvania State University, University Park, PA, 16802, USA
e-mail: xiliu@psu.edu; maskedchicken@gmail.com; clio@psu.edu

L. Perez et al. (eds.), *Agent-Based Models and Complexity Science in the Age of Geospatial Big Data*, Advances in Geographic Information Science,
DOI 10.1007/978-3-319-65993-0_6

production (a static representation), conceptualizing their position within an urban system such as the hierarchy (using a dynamic representation of in and out flow) allows researchers to examine the city within this larger corpus of transactions.

The urban hierarchy is comprised of groups of spatial regions where each is anchored by one very large city. This large anchor city (e.g. Chicago) exerts a gravitational pull on its surrounding cities unless another large city, perhaps Minneapolis, MN or St Louis, MO claims what would usually be Chicago's surrounding cities as part of their own functional regions. Cities that lie on a region's periphery or circumference are more likely to switch regions than those closer to the anchor city. In the past, regions were bound into geographically-cohesive areas in order to minimize the costly movement of natural resources and commodities [1]. When peripheral settlements send many flows to a city's central business district, this settlement is considered part of the larger city's functional region. In practice, flows such as migrants and commuters help the U.S. Office of Budget and Management define the spatial boundaries for Business Economic Areas (BEAs) and Metropolitan Statistical Areas (MSAs).

The traditional regional approach has been explained by the gravity model, which estimates interaction (i.e. flows, connection strength) between two places as the product of their respective populations divided by the square of the distance between the places [2]. The resulting estimate simulates the economic pull strength of cities, assuming that many people will choose to connect to a nearby place whose large size signifies many opportunities [3]. A gravity model using just population and distance has been shown to predict about 57% of U.S. inter-city migrant flows [4].

Regions are also delineated by areas of homogenous industry [5] or cohesive economic activity [6]. More recently, creative methods like dollar bill circulation [7], telephone calls [8], surname clustering [9] and maps of sports team popularity from Facebook likes [10] have been used to delineate regions around functional anchor cities.

Today, the regional hierarchy can be re-examined with a network approach. The economic transition from manufacturing to digital services and information technologies has allowed regions to form and function not just as a group of nearby cities, but as a network of connected cities that may or may not be proximal. This network is formed by "leapfrogging" (skipping over) nearby cities to create connections with distant cities that have economic benefit [11, 12]. These networked economies are not geometrically contiguous and thus, the connections are harder to predict in theory, but larger and more comprehensive data sets allow for the investigation of factors beyond traditional place-to-place connectivity (such as the gravity model) [13].

Here, we focus on descriptive properties of migration in the U.S. urban system. Migration choice has been explained by factors such as searching for the best job possible [14] seeking out a certain lifestyle [15], or capitalizing on social networks and interpersonal relationships [16]. Instead of building a model that attempts to correlate high migration volume with demographic or economic variables of different cities, we view the migration system as a network of cities connected by volumes of migrants, as per the topic of migration systems theory (MST) [17]. Similar studies partition city systems into communities that are closely connected internally [18, 19], or search for network hub cities [20, 21]. These studies advance the use of network science in studies of the urban hierarchy, but do not address the extent to which migrants surpass near cities to connect with those further away—as they may when interpersonal relationships and institutions are involved.

We use county-to-county migrant flows sourced from the U.S. Internal Revenue Service (IRS) for 21 years (details in Sect. 2) to form a network of 917 cities (nodes) that connect to each other (edges) weighted by the number of migrants exchanged by cities (an undirected network). Because the complete network of migrants ties many cities together and we are interested in uncovering the urban hierarchy, we experiment with the following concepts:

1. *Best Friend*: Best friend networks are created by drawing an edge between an origin and the destination to which it most frequently sends migrants.
2. *Best High/Low Income Friends*: These networks differentiate high-income flows from low-income flows. The network is made from gathering each city's highest income outflow and connecting it with that destination. (i.e. an edge is made to the destination that attracts migrants with the highest average income). The same procedure is repeated for each city's lowest income destination.
3. *Constellations*: This method produces a collection of graphs (i.e. disconnected subgraphs of networks) of cities that due to the number of nodes involved in each graph, their configuration and their spatial genesis, resemble constellations which can be classified into motifs. We create a single 'galaxy' of constellations for each of 21 years.

Our results show that migration networks exhibit significant structural temporal persistence, and clear ensemble rules can be used to construct the networks. We find that some cities switch preferences to alternative large city anchors over time, and that some large city anchors become popular or decline in popularity. We also determine that low-income flows create different networks than high-income flows. We validate and contextualize these findings by comparing our model to the gravity model and radiation model. Our proposed networks can respond to the following questions: Which cities are popular for migrants? What regions (i.e. connected graph structures) arise? Which cities feed into larger cities? Which cities bypass closer and larger cities to connect directly to a more distant metropolis? Does a population hierarchy emerge? Are systems of cities closed or do they connect in larger chains? How do these patterns change for high- and low-income migrants?

In Sect. 2, we describe the migration dataset, network and analysis methods. In Sect. 3, we explore re-occurring constellations in the networks, compare our model to other prevailing models such as the gravity and radiation model, which reflect the structure of urban hierarchy. We conclude in Sect. 4.

2 Data and Methods

2.1 *U.S. Migration Data and Population Data*

We use data from the U.S. Internal Revenue Service (IRS) Statistics of Income Migration Data for years from 1992–1993 through 2012–2013 for this study. These data are free and available online. The original data were generated from the yearly change in address reported on individual tax returns from one year to the next, and aggregated at the county level to produce a network of county-to-county flows. Each flow contains three attributes: the number of returns, the number of exemptions, and the adjusted gross income (AGI), which is the sum of all income moving on the flow. Flows must contain at least ten returns to be reported in the dataset. We use number of exemptions to estimate the migrant population, as this value reflects the size of families, including children and jointly-filing spouses. Alternatively, using the number of filers would estimate the number of heads of households that migrate.

We aggregated the county-to-county flows into flows among Core Based Statistical Areas (CBSAs), formerly referred to as MSAs. CBSAs are defined as urban cores and peripheries with a population of at least 10,000 residents. Since CBSAs (henceforth, cities) follow county boundaries, aggregation required only flow summation. The aggregated data contains 917 cities reporting migration flows throughout the 21 years period. Each city is accompanied by a population count defined by the U.S. Census Bureau at the county level, as aggregated to the city level.

2.2 *The Best Friend Configuration Model*

The network is configured based on the *single allocation* [22] of edges to nodes (i.e. cities). In this configuration, a single city is only permitted to attach to the city to which it sends the highest proportion of its flows. For example, New York City is only attached to Miami because it sends more migrants to Miami than to any other city. In this model, each city is allowed only one outgoing connection (out-degree = 1), but the in-degree can be as large as the number of other nodes in the system (n−1). Thus, this network is referred to as the *best* friend network. Edges are assigned a weight (w) calculated as the proportion of migrants (m) city *i* sends to city *j* (Eq. 1):

$$w = \frac{m_{ij}}{\sum_1^k m_{ij}} \tag{1}$$

where *k* is the total number of cities to which city *i* is connected (i.e. its outgoing degree).

Over 21 years, the best friend model contains a total of 3.1% possible city-to-city edges, but accounts for 20.7% of total system-wide migrants. The average yearly migration flow magnitude ranges from 170–200 migrants and the average best friend magnitude ranges from 1100–1500 migrants. Over time, the total number of system-wide migrants grew from 5 million to 6.6 million and best friends accounted for 1 to 1.4 million migrants each year (hence, about 20% of total migration). Average AGI incomes range from $120,000 to $9000 per flow.

We also derive two special types of best friend models where edges are characterized by average income on the flow, calculated as the AGI of that flow divided by the number of returns on the flow. High and low-income migration networks are each created by selecting the best-high-income friend and best-low-income friend of a city, defined as the largest migration streams amongst the top 10% (high-income) and bottom 10% (low income) average income migration connections leaving city i. The top 10% is used rather than single the highest/lowest income flow to ensure a high number of migrants and thwart anomalies.

Temporally, each city in the income networks has an average of nine different best-high/low-income friends over the time period. On average, cities are connected to their best-high-income friends for 6.5 years and to their best-low-income friends for 5.4 years. Generally, wealthy best friend pairs are more stable over time.

2.3 Constellations

Constellations, or motifs [23], are basic graph structures that repeatedly appear in networks. In this study, constellation is a relaxed definition of motif that refers to families of basic structures that are widely seen in best friend networks, as compared to their probability of arising in a null models based on the gravity model. We detected five types of motifs (Fig. 2) in the best friend networks: pairs, chains, hubs, stars, and trees [24]. These graph structures are enumerated and analyzed using community detection methods within the R statistical computing environment's *igraph* package [25, 26]. Their definitions are as follows:

Pairs: A pair is formed by two cities that are each other's best friend. They are isolated from population hubs and may have strong dependency on each other.

Chains: A chain is a series of single directional connected cities where for $i = 1 \ldots n$, city i points to city $i + 1$. Usually, city n connects to a local hub. Chains can reveal how a series of many migrants connect to nearby non-hub cities, possibly facilitated by a lack of social connections in large cities, poor mobility, or high levels of local social capital.

Hubs: A hub is a node with an in-degree larger than one with 'spoke' cities directly connected to it. The hub node may point to one of its spokes or to other hubs, creating stars and trees. Hubs are popular destinations for both chain and non-chain nodes.

Stars: A star is defined by hubs that point to one of their spokes. Small local hubs tend to form stars with proximal cities, and rely less on the influence of distant, larger hubs.

Trees: A tree is hub that connects to other hub nodes, and is typically attracted by higher-level hubs. Trees tend to connect small, medium and large cities.

2.4 Analytical Methods

Gravity model. The gravity model, as in [27], is a classical model for predicting flows based on population and distance, so that the magnitude of migrants T_{ij} between city i and j is estimated as:

$$T_{ij} = K\frac{P_i P_j}{d^{\beta}} \tag{2}$$

where P_i and P_j represents the population in city i and j, respectively, d is the distance between the two cities and K is a constant. β is a distance decay factor, often referred to as the coefficient of friction, and most commonly estimated with value of 2.

Radiation model. The radiation model [28] is used to predict flow volumes T_{ij} between city i (with population P_i) and j (with population P_j) as:

$$T_{ij} = T_i \cdot \frac{P_i P_j}{\left(P_i + s_{ij}\right)\left(P_i + P_j + s_{ij}\right)} \tag{3}$$

where T_i represents outflows from city i, and s_{ij} denotes the total population of alternative population centers within a given radius of the destination city.

Distance between two cities is calculated as the using Euclidean distance between each CBSA's geometric centroid.

3 Results

3.1 Best Friend Network

A series of best friend networks was created for each year (ex. Fig. 1). Most cities (60%) have no in-degree (degree = 1), 21% of cities have a degree of two (one outgoing flow, one incoming flow), 9% have a degree of three and 5% of cities have a degree of six or higher. These larger hubs include the U.S.'s ten largest cities, with Dallas consistently having the highest degree at over 20 best friend connections. The best friend network detects the regional importance of more geographically-

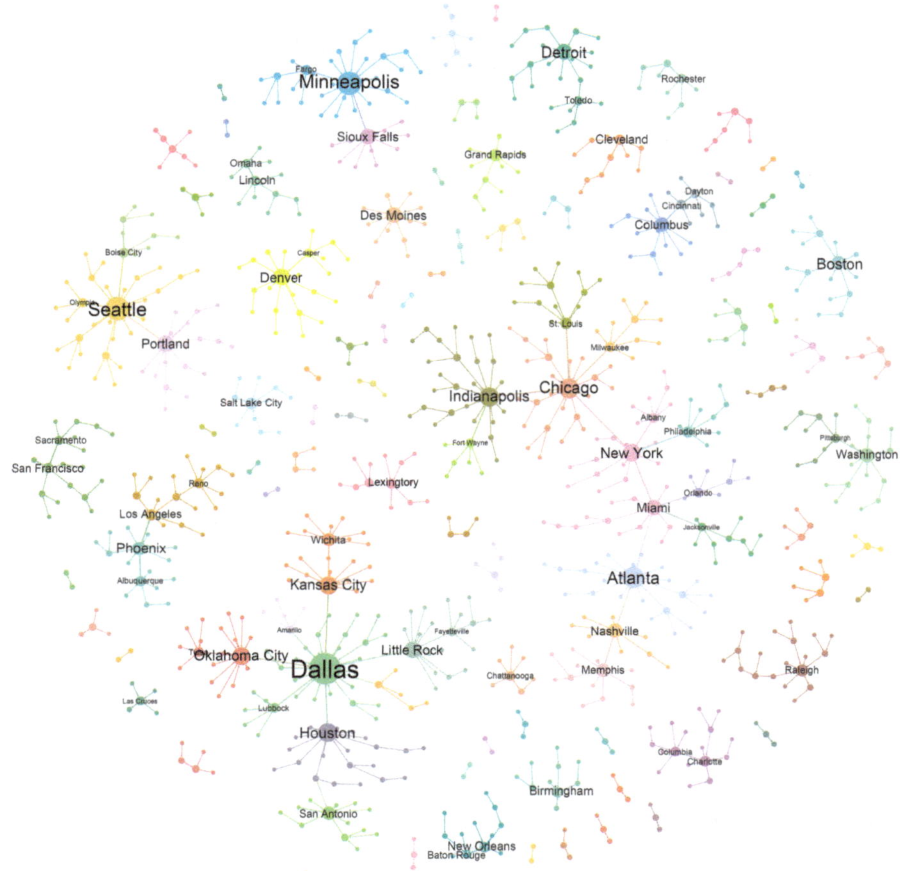

Fig. 1 Best friend network constellations for year 2012. In the network, the size of nodes corresponds to their degrees using Yifan-Hu's proportional method in Gephi [29] and each color represents a separate constellation. Some connected constellations are divided into different components due to their relatively weak connections, as determined by the community detection algorithm [25]

isolated cities such as Oklahoma City, Sioux Falls, Salt Lake City, Wichita, Des Moines, Memphis, Jackson, Grand Rapids, and Little Rock (Fig. 1) in their local hierarchical systems.

As migrant streams change each year, we can expect some fluctuation in the network. The average time spent with a best friend is 13.8 years. 504 of 917 cities (55%) have only one best friend for the entire period and 877 (96%) have at most three different best friends. On average, 110 cities change best friends each year, a turnover rate of 12% per year. Because the number of different best friends is low, this turnover rate does not compound at a high rate over longer time periods (e.g. 15% of cities have a different best friend in 2000 than in 2012). The strength of a best friendship, i.e. the percentage of migrants sent to a best friend city (Eq. 1), ranges from 3.6% (Chicago to Los Angeles in 1992) to 100% for Mount Sterling, KY to Lexington, KY (1996–1999; 2001) and Big Stone Gap, VA to Kingsport-Bristol, TN-VA (2006). In general, cities Chicago, Columbus, OH and Atlanta have the smallest percentages of migrants sent to their best friend cities.

Crucially, we do not see an increase in the diversity of places to which a city sends its migrants. We had hypothesized that the rise of the Internet and mobile technologies in the late 1990s would promote more swirling/churn in the preferences of the migrants, given the new opportunities to research potential destinations. With more diverse information, migrants may have experienced other places, i.e. travelled more, and garnered friends in multiple locales. Yet, our analysis does not reveal a diversification of movement over multiple destinations at any point during this time. In fact, we see a steady increase in the average percentage of migrants a city sends to its best friend, starting at 0.35 in the early 1990s and rising to 0.37 in the 2010s.

The number of separate constellations and their size remains relatively stable over time. For each year, there was an average of 105 constellations, each containing from 2 to 66 cities, with an average size of 8.8 cities. The majority of constellations are small clusters, with 80% of the constellations comprised of fewer than 13 cities and 49% comprised of fewer than 5 cities (Fig. 1). Most constellations are geographically compact (averaging about 130 km), driven in part by small constellations, especially mutual best friend pairs which limits the average geographic spread. Notable exceptions include the strong New York City-Miami connection and any constellation connecting Alaska or Hawaii to the mainland, as well as some recurring connections between major cities. The New York City-Miami connection is driven in part by retirees from New York City choosing to move to a warmer climate (Fig. 1). These migrants are colloquially known as "snowbirds".

We next categorize individual nodes based on the following observed motifs: pairs (both isolated pairs and those pairs within larger constellations), hubs (star and non-star hubs), spokes (nodes directly connect to hubs and with 0 in-degree), two categories of trees: {tree hubs (local hubs in a tree motif) and tree spokes (nodes directly connect to tree hubs and with 0 in-degree)}, and chains (all members in a chain category) (Fig. 2, Table 1). A node can only be placed into one category. The categorization is implemented with an algorithm that uses in/out degree, the in/out degree of their best friend, and the node type of their best friend as input parameters. We find that most constellations, especially large constellations, have single central nodes. Since large hubs have more resources, they may be less likely to rely on other hubs.

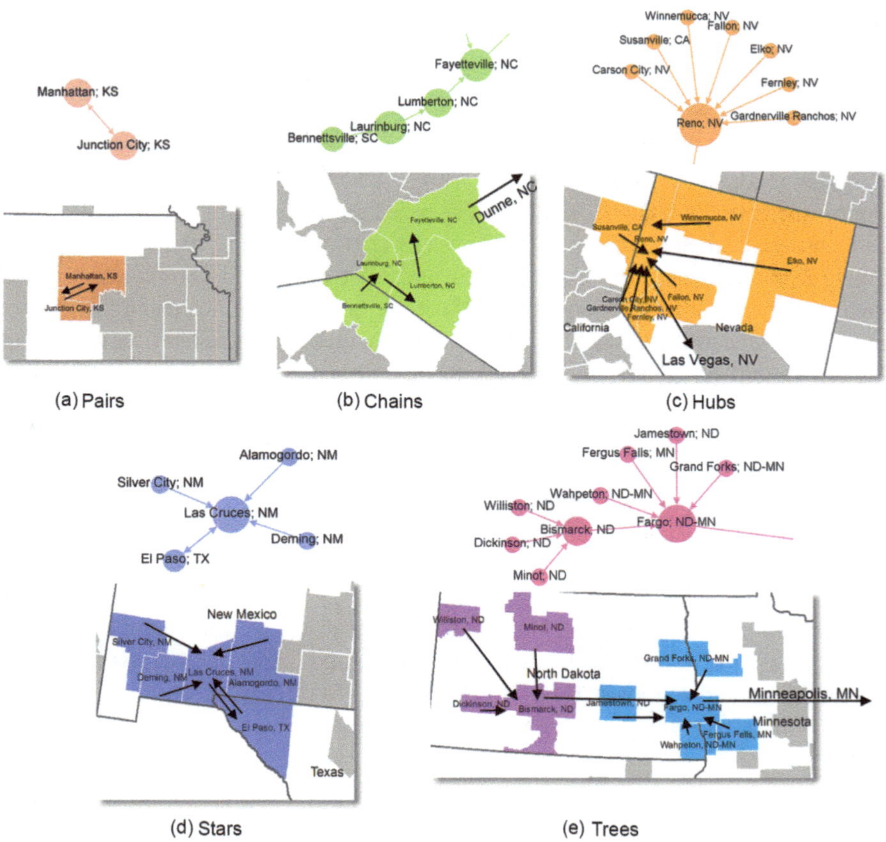

Fig. 2 Graph motifs with geographic examples. When cities are connected to their best friends, different network motifs arise, including pairs, chains, hubs, stars, and trees. These schematics illustrate differences between the roles of distinct cities within their regional systems, and what the regional systems look like as a network of flows

Table 1 Proportions of different types of nodes throughout 21 years in best friend networks

Node types percentage (%)	Hubs	Spokes	Pairs	Tree hubs	Tree spokes	Chains
Best friend	12.1	30.2	8.3	7.3	19.3	22.8
High-income	4.4	20.2	0.9	12.4	26.9	35.2
Low-income	6.5	13.4	2.7	13.9	26.4	37.2

3.2 Comparison to Prevailing Methods

The best friend method highlights the backbone structure of urban hierarchy. Since the structure is based on migration flows, we compare the best friend method with related prevailing models, such as the gravity model and radiation model, to demonstrate that the urban hierarchy produces some patterns that are not well explained by population and distance.

Using the most recent data, we find that a parameterized gravity model with a pre-determined coefficient of friction (β) of 3 predicts about 60% of best friend (n = 552). The closest city is a city's best friend 28% of the time (n = 259). The radiation model predicts the best friend with 39% accuracy (n = 357). We visualized the best friend network in 2012 (Fig. 3a) based on a fitted gravity model for each city and the corresponding constellations. Instead of showing various motifs, the network is dominated by hub-spoke structure. If two cities do not connect in the data, we do not consider them as candidates for best friend cities using the gravity model. In other words, cities may have a clear choice given the gravity model, but if the city did not send any migrants to this attractive choice (or choices), they connected to their next best choices to which they actually sent migrants.

We also discover cities that "defy" gravity. We normalized the flow weight (w) by the interaction measured by a fitted gravity model to isolate flows that are large despite a small interaction estimate. These cities draw origin cities despite high travel cost (distance) and relatively low population. These networks contain more long-distance connections than best friend networks, a manifestation of "leapfrogging" in the hierarchy, and illustrate how migration connects labor forces to employment opportunities. For example, San Jose, in Silicon Valley, has an agglomeration of leading technology companies, and connects to faraway college towns throughout the years (Fig. 3b); Williston, North Dakota, becomes a hub in the network after oil resources were found in the early 2010s, and spurred jobs and economic growth.

3.3 High- and Low-Income Routing

The motifs of high and low income networks are collectively distinguished from best-friend networks. First, there are fewer pairs in the income networks, presumably because smaller towns that depend on each other in general take more preferential (and less mutual) routes when income is involved. Interestingly, there are very few pairs that exchange high-income migrants (0.9%), but three times as many will be best friends for low-income migrants (2.7%) (Table 1). There are more chains in the income networks although these chains are shorter. There are more hubs that are parts of trees in the income networks, indicating that hubs also connect to other hubs in these networks (Table 1).

There are also differences between low-income and high-income networks. High income networks have fewer hubs, but these hubs are quite large, as indicated by in-degree, and draw more distant connections (Fig. 4a), while hubs of low-income networks have fewer spokes. The high-income network has 39.38 unique constellations and the low-income network has 51.14 constellations, indicating more local regionalization and less overall connectivity in the low-income network. The

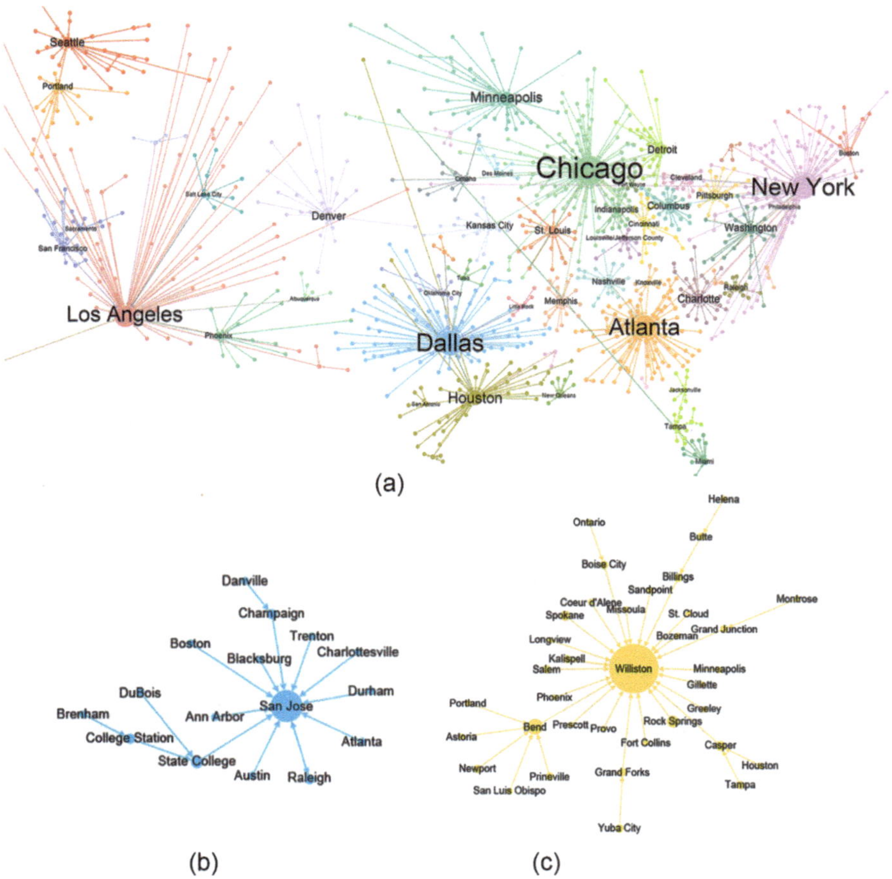

Fig. 3 (**a**) Best friend network in 2012 with best friend cities selected based on a fitted gravity model. The network is dominated by hub-spoke structure and real world motifs such as trees and chains are rarely seen. However, our data reveal that best friends do not always choose the destination predicted by the gravity model. For example, San Jose, California (**b**) is the hub for many college towns (this example is derived from the 2010 annual dataset), while Williston, North Dakota (**c**) starts attracting many cities after a boom in the energy industry (this example is derived from the 2012 annual data)

average constellation size is 23.49 nodes for the high income network and 18 nodes for the low-income network (Fig. 4b). The temporal change of constellation number and size does not have significant trends; the average distance of best low-income friends fluctuates while the average distance between high-income friends grows, suggesting greater mobility (Fig. 4a).

When mapping the constellations onto the geographic boundaries of their respective cities, we find that the low-income network depicts more local clustering, i.e. cities within the same constellation tend to be nearby, and also tends to follow state lines (Fig. 4c, d). Conversely, the high-income network constellations are not

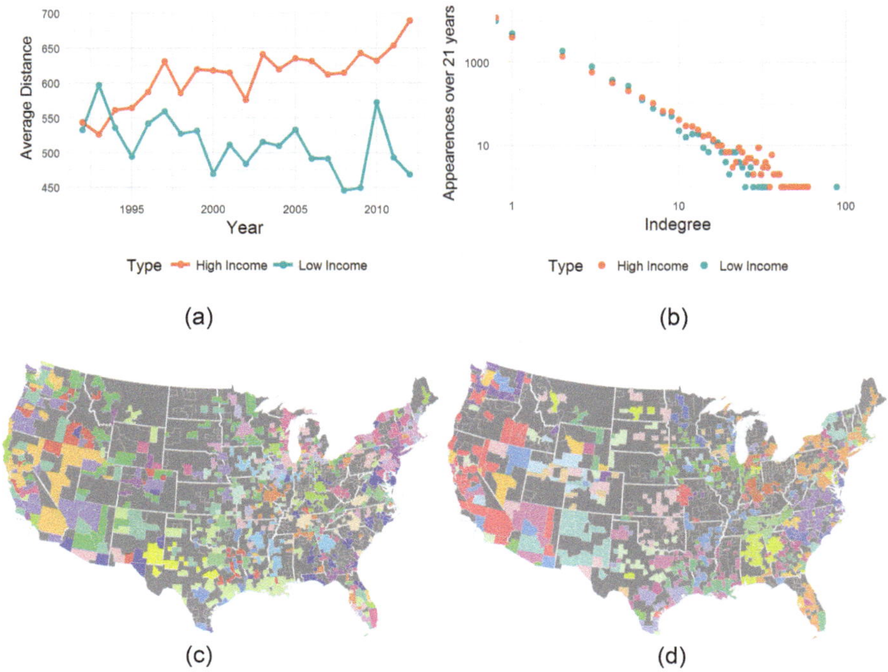

Fig. 4 High and low income network results. (**a**) Average edge distance in the high-income network grows over time, while distances between low-income cities have no clear trend. (**b**) The in-degree distribution of nodes in best-high/low-income-friend networks throughout the 21 years show that the high-income network formed hubs with high degrees, and the low-income network is marked with more cities with few incoming flows. The top 20 largest constellations using the 2012 data were identified and mapped to their corresponding city in high-income (**c**) and low-income (**d**) networks. Cities with the same colors belong to the same constellations

as contained geographically; cities in many different states often belong to the same cluster, indicating that neither boundaries nor distance appear to deter movement as significantly as in the low-income network (Fig. 4c, d).

4 Conclusions

In this study, we proposed a series of methods to study the U.S. urban hierarchy using 21 years of migration data from the U.S. IRS. We built a single allocation (best friend) network from all migrant flows, and similar networks highlighting only high- and low-income flows. Our results showed that the best friend network did not align well with the gravity and radiation models of urban interaction, and was distinguished by urban hubs, spokes and chains. Cities also tended to keep a maximum of three best friends over the time period. The income networks were marked with stronger hubs (with more spokes), that served in a system of connected hubs. The high-income network encouraged longer flows, more leapfrogging, and exhibited more spokes that directly migrate to hubs that attract high earners nationwide, such as Cape Coral and Naples, Florida. The low-income networks contain a few more chains, which may represent that low-income migrants are more likely to move to nearby cities first and 'climb up' to the hubs gradually, which may result from limited mobility and limited social capital in hub cities.

The biggest limitation of our study is the variation in the meaning of a city's "best friend". For some cities, the best friend is a significant dependent tie as a city may send all of its migrants to this city, while for others (such as Chicago and Memphis), the best friend only absorbs about 5% of migrants. The edge that results from both of these scenarios is indistinguishable in the network. One potential remedy is to use analytical methods that account for edge weights.

The larger, eventual goals of testing the best friends method is to use it to (1) unearth ties that may not make sense economically, but may be the result of interpersonal relationships, and (2) come closer to understanding how flows affect the places to which they connect. When a city's migrants are attracted to a city, we consider these cities to be in a similar functional region—as they are exchanging the same people between multiple cities. These functional regions are increasingly geographically disconnected, which should be accounted for in geographic partitioning exercises and in location-allocation models that use distance as an input parameter.

Acknowledgments Thanks to Prathamesh Aher for helping prepare data.

References

1. Greenwood MJ (1985) Human migration: theory, models, and empirical studies. J Reg Sci 25:521–544
2. Dodd SC (1950) A gravity model fitting physical masses and human groups. Am Sociol Rev 15:245–256
3. Pred A (1980) Urban growth and city systems in the United States, 1840–1860. Harvard University Press, Cambridge, MA
4. Andris C, Halverson S, Hardisty F (2011) Predicting migration system dynamics with conditional and posterior probabilities. In: ICSDM 2011—Proceedings 2011 IEEE international conference on spatial data mining and geographical knowledge services
5. Harris JR, Todaro MP (1970) Migration, unemployment and development: a two-sector analysis. Am Econ Rev 60:126–142
6. Green HL (1955) Hinterland boundaries of New York City and Boston in Southern New England. Econ Geogr 31:283–300
7. Brockmann D, Hufnagel L, Geisel T (2006) The scaling laws of human travel. Nature 439:462–465
8. Calabrese F, Dahlem D, Gerber A, Paul D, Chen X, Rowland J, Rath C, Ratti C (2011) The connected states of America: quantifying social radii of influence. In: Privacy, security, risk and trust (PASSAT) and 2011 IEEE third inernational conference on social computing (socialcom). pp 223–230
9. Cheshire JA, Longley PA, Yano K, Nakaya T (2014) Japanese surname regions. Pap Reg Sci 93:539–555

10. Meyer R (2014) The Geography of NFL Fandom. The Atlantic. https://www.theatlantic.com/technology/archive/2014/09/the-geography-of-nfl-fandom/379729/
11. Kotkin J (2001) The new geography: how the digital revolution is reshaping the American landscape. Random House, New York
12. Heim CE (2001) Leapfrogging, urban sprawl, and growth management: Phoenix, 1950–2000. Am J Econ Sociol 60:245–283
13. Treyz GI, Rickman DS, Hunt GL, Greenwood MJ (1993) The dynamics of U.S. internal migration. Rev Econ Stat 75:209–214
14. Greenwood MJ, Sweetland D (1972) The determinants of migration between standard metropolitan statistical areas. Demography 9:665–681
15. Chen Y, Rosenthal SS (2008) Local amenities and life-cycle migration: do people move for jobs or fun? J Urban Econ 64:519–537
16. Andris C (2016) Integrating social network data into GISystems. Int J Geogr Inf Sci 30:2009–2031
17. Fawcett JT (1989) Networks, linkages, and migration systems. Int Migr Rev 23:671–680
18. Thiemann C, Theis F, Grady D, Brune R, Brockmann D (2010) The structure of borders in a small world. PLoS One 5:e15422
19. Manduca RA (2014) Domestic migration networks in the United States. Massachusetts Institute of Technology, Cambridge, Massachusetts
20. Guimera R, Mossa S, Turtschi A, Amaral LAN (2005) The worldwide air transportation network: anomalous centrality, community structure, and cities' global roles. Proc Natl Acad Sci, USA 102:7794–7799
21. Davis KF, D'Odorico P, Laio F, Ridolfi L (2013) Global spatio-temporal patterns in human migration: a complex network perspective. PLoS One 8:e53723
22. O'Kelly ME (1998) A geographer's analysis of hub-and-spoke networks. J Transp Geogr 6:171–186
23. Dunne C, Shneiderman B (2013) Motif simplification: improving network visualization readability with fan, connector, and clique glyphs. In: Proceedings of the SIGCHI conference on human factors in computing systems. pp 3247–3256
24. Jackson MO (2008) Social and economic networks. Princeton University Press, Princeton, N.J
25. Clauset A, Newman MEJ, Moore C (2004) Finding community structure in very large networks. Phys Rev E 70:66111
26. Csardi G, Nepusz T (2006) The igraph software package for complex network research. Int J Complex Syst 1695:1–9
27. Tarver JD, McLeod RD (1973) A test and modification of Zipf's hypothesis for predicting interstate migration. Demography 10:259–275
28. Simini F, González MC, Maritan A, Barabási AL (2012) A universal model for mobility and migration patterns. Nature 484:96–100
29. Bastian M, Heymann S, Jacomy M (2009) Gephi: an open source software for exploring and manipulating networks. In: Third International AAAI Conference on Weblogs and Social Media. pp 361–362

Discovering Multi-Scale Community Structures from the Interpersonal Communication Network on Twitter

Caglar Koylu

Abstract Despite the controversies of privacy and ethics, spatially-embedded communication data from widespread and emerging online social networks provide an unprecedented opportunity to study human interactions at the global scale. Detecting communities of individuals who live close by and have strong communication among each other is critical for a variety of application areas such as managing disaster response, controlling disease spread, and developing sustainable urban spaces and infrastructure. The ease of long-distance travel and communication have generated a highly complex network of human interactions, in which long-distance and short-distance ties coexist in multiple scales. Also, there is a hierarchical spatial organization in human interaction networks which reflect historic and socio-political borders. Patterns of human connectivity cross these historic and socio-political borders at multiple geographic scales. Therefore, a comprehensive understanding of human interactions necessitates analysis methods to take into account the complexity introduced by the multi-scale nature of human connectivity. This paper employs a spatially-constrained hierarchical regionalization algorithm to reveal multi-scale community structures in the interpersonal communication network on Twitter. The interpersonal communication network was constructed using a year of reciprocal and geo-located mention tweets in the U.S. between August 2015 and 2016. The results strikingly showed nested borders of cohesive regions at multiple scales, which are inherent to human communication patterns in the regional hierarchy of the U.S. Unsurprisingly, people communicated with others that live nearby, and multi-scale regions overlap with administrative boundaries of the states, cultural and dialectal regions, and topographical features. Furthermore, visualization of interregional communication patterns revealed a variety of spatial connectivity patterns such as poly-centricity, hierarchies, and spanning trees. Discovery of such patterns is essential for understanding of the complex social system that is influenced by long-distance ties.

C. Koylu (✉)
Department of Geographical and Sustainability Sciences, University of Iowa, 316 Jessup Hall, Iowa City, IA, 52242, USA
e-mail: caglar-koylu@uiowa.edu

L. Perez et al. (eds.), *Agent-Based Models and Complexity Science in the Age of Geospatial Big Data*, Advances in Geographic Information Science,
DOI 10.1007/978-3-319-65993-0_7

Keywords Community detection • Hierarchical regionalization • Interpersonal communication • Twitter mentions • Geo-social networks

1 Introduction

Despite the controversies of privacy and ethics, in recent years, publicly available data from location-based social networks (LBSN) such as Twitter, Foursquare, Gowalla, and BrightKite have made it possible, for the first time in human history, to examine human interactions at the global scale. One can infer human interactions through various forms of geo-tagged communication data such as text, photo, video, and check-in locations provided by online platforms. Understanding of human communication and social ties is crucial for addressing societal challenges such as managing disaster response, controlling disease spread, and developing sustainable urban spaces and infrastructure.

Previous studies in LBSN have utilized various forms of communication data to analyze the effect of geographic proximity on social interactions [1–3]; and the structural and geographic characteristics of communication networks at the global scale [4–8]. In addition to understanding global characteristics of communication networks, there has been a growing interest in identifying community structures in human mobility and communication networks [9–11]. Findings of these studies across various themes highlight strong resemblance of human communication and mobility patterns, and the constraining effect of administrative boundaries, topographical features, cultural and linguistic variations on human mobility and communication [12]. However, the ease of long-distance travel and communication have generated a highly complex network of human interactions, in which long-distance and short-distance ties coexist in multiple scales. Also, there is a hierarchical spatial organization in human interaction networks which reflect historic, and socio-political borders. Patterns of human connectivity cross these historic and socio-political borders at multiple geographic scales [9, 10, 13–15]. Therefore, a comprehensive understanding of human interactions necessitates methods that take into account the complexity introduced by the multi-scale nature of human connectivity.

This paper employs a spatially-constrained hierarchical regionalization algorithm to reveal multi-scale community structures in the interpersonal communication network on Twitter. The interpersonal communication network was constructed using a year of reciprocal and geo-located mention tweets in the U.S. between Aug. 2015 and 2016. The results strikingly showed nested borders of cohesive regions at multiple scales, which are inherent to human communication patterns in the regional hierarchy of the U.S. Unsurprisingly, people communicated with others that live nearby, and multi-scale regions overlap with administrative boundaries of the states, cultural and dialectal regions, and topographical features. Furthermore, visualization of interregional communication patterns revealed a variety of spatial connectivity patterns such as poly-centricity, hierarchies, and spanning trees.

2 Related Work

2.1 Distance and Social Interactions

Social ties and communication are constrained by distance, and most of them are geographically local [4]. Deville et al. [16] have shown a great similarity between communication and mobility patterns, and explain the spatial dependencies by a scaling relationship using power laws. Similarly, Emmerich et al. [17] analyzed a variety of spatially-embedded networks such as the Internet, power grid, transportation and communication networks, and found that spatial constraints are relevant, and the relationship between topological and geographic distance varies by dimension and scaling factors. Von Landesberger et al. [18] introduced a flow clustering and visualization approach to identify spatiotemporal variation in the mobility and communication patterns from tweets and phone call records. Von Landesberger et al. [18] found similarities in spatiotemporal patterns such as movements and communication directed from/to central locations given a particular cycle (e.g., daily, weekly). McGee et al. [19] analyzed the effect of distance on the strength of ties, and classified Twitter's utility both as a social network of geographically nearby friends, and as a news distribution network of individuals that live far apart. Higher intensity of communication has also been found to be associated with external factors such as gender, demographics, and socio-economic status. By analyzing 30 billion online conversations, Leskovec and Horvitz [6] found that people tend to communicate more with each other when they have similar age, language, and location; and cross-gender conversations are both more frequent and of longer duration than conversations with the same gender.

Different forms of communication data have been analyzed to examine geographic and structural characteristics of human communication. Krings et al. [20] and Lambiotte et al. [21] revealed that the communication intensity between two cities can be estimated as a function of population, distance, and predominant language using phone call records. Barnett et al. [22] also analyzed phone call records and found that the relationship between homophilly and spatial autocorrelation is amplified in places with high density of individuals. Garcia-Gavilanes et al. [23] studied Twitter user mention network, and found that the probability of two user mentioning each other correlates with power distance. Several studies [24–26] have shown similar findings, and revealed that user mentions on Twitter occur between users that are in close geographic proximity. In addition to distance, Garcia-Gavilanes et al. [23] incorporated economic, cultural and social variables to predict the volume of communication flows between countries. Herdagdelen et al. [27] analyzed social, political and geographic characteristics of news-sharing communities on Twitter, and defined social groups based on local, national and global level. By analyzing a large Twitter dataset, Groh et al. [28] found that (1) the social tie strength decreases as expected with increasing spatial distance among users (2) the information value decreases when the tie strength increases; and (3) the value of information is independent from the distance.

2.2 *Community Structures in Spatial Networks*

In a network, a community is defined as a set of nodes (individuals) in which the density of connections is stronger internally within the community than it is externally with the individuals from different communities [29]. Community detection algorithms without explicit spatial constraints [30] can be applied to identify communities in spatial networks, which may be multi-part (split) in geography. Various modularity-based community detection algorithms have been used to discover community structures in networks of human mobility [31], commuting [32], telephone call records [10], friendship networks [11], twitter [33, 34], and credit card transactions [35]. Communities discovered by these studies are often geographically confined to nearby regions, however, some of them are multi-part in geographic space. To bridge the geographical and network aspects of communities, Croitoru et al. [36] integrated Louvain and density clustering methods to identify and link community structures in the network (cyber) space and geographic space. Similarly, gravity models have been applied in non-spatial and modularity-based community detection algorithms [37] to estimate expected flows as a function of geographic distance, and derive geographically cohesive community structures. Alternatively, one can embed spatial constraints in community detection to partition a spatial network into smaller sets of contiguous nodes or functional regions that are densely connected internally. In this paper, a spatially-constrained hierarchical regionalization algorithm [9] is used to reveal multi-scale community structures in the spatially-embedded reciprocal mention network.

3 Data and Network Extraction

Geo-located tweets in the Contiguous U.S. between Aug. 1, 2015 and Aug. 1, 2016 were collected using the Twitter Streaming API. Location of tweets are available in two different levels of granularity: exact geographic coordinates, or in a descriptive manner by listing of a place name such as a city. Stefanidis [38] reported that 0.5 and 3% of the tweets had precise coordinates over a period of two years prior to 2013, and also highlighted that the use of precise coordinates increased to 16% during events such as Fukushima disaster in Japan. The dataset used in this paper included 14% of the tweets with precise geographic coordinates, which could potentially be attributed to increasing adoption of mobile technology. In this paper, tweets with both exact geographic coordinates and place names that corresponded to an area at least at city scale were used. Therefore, place names that were at the state or country level, which corresponded to 18% of the tweets with place names, were excluded. As a result, the dataset of tweets with exact coordinates and place names that are at least at city level, consisted of 700 million tweets, and 6.6 million users.

Communication between Twitter users is handled through a set of functions. Follower, favorite and retweet functions are useful for modeling information diffusion, whereas mentions and replies allow users to join conversations on Twitter, wherein direct personal communication could be extracted [36]. A reply is a response to another user's tweet that begins with the @username of the original poster, a mention is a tweet that contains another user's @username anywhere in the body of the message. In a user mention, the tweet includes only the location of the sender who mentions another user (recipient), and a representative location of the recipient in a mention can be derived only if the recipient has at least one geo-located tweet in the sample. Also, since individuals are mobile, locations of tweets from each user are variable across space. In this paper, tweet locations were overlaid with census data (e.g., county boundaries) to identify a home area for each user based on the most frequent tweet location. Another commonly used strategy could be to determine the home location based on tweets posted at night time where individuals are assumed to be home. In this paper, only the reciprocal mention pairs, or in other words, back-and-forth conversations [37] were used while the tweets that were not replied were disregarded.

A data cleaning procedure was performed prior to constructing the geo-located user mention network on Twitter. Using the metadata provided by the Twitter Streaming API, the following tweets and users were filtered out: (1) the tweets authored by non-personal user accounts such as news feeds, weather and emergency reports, and external applications such as Foursquare and Instagram (2) users with more than 3000 followers to prevent any bias caused by a large number of user mentions attracted by a few users, i.e., celebrities [39]. After the cleaning process, the number of tweets decreased to 290 million (42%). Of these 290 million tweets, 221 million (76%) included a user mention. There were 4.7 million users who were mentioned in a tweet at least once.

After the initial data cleaning, the following steps were performed to extract the reciprocal mention network. First, a spatially embedded individual-to-individual reciprocal mention network was constructed by taking into account the tweets of users who both send and receive messages between each other. Of the 221 million mention tweets, 71 million tweets (32%) corresponded to tweets exchanged between users that both users' home county can be located. After further filtering to obtain reciprocal mentions, the number of tweets was reduced to 33 million (46% of geo-located mentions). The individual reciprocal pairs were then aggregated into a county-to-county network by using the most frequent county location for each user. In the county-to-county network, a link illustrates the total number of reciprocal pairs between two counties.

4 Methodology

A spatially-constrained hierarchical regionalization algorithm [9] was employed to reveal multi-scale community structures in the spatially-embedded reciprocal mention network. The regionalization method produces a hierarchy of spatially contiguous regions, where there are more flows within regions than across regions.

First, a modularity measure of connection strength was computed rather than using the raw flow counts (reciprocal pairs) between each pair of locations. This step is necessary to remove the effect of population by calculating the difference between the actual flow and the expected volume of flow for each pair of locations (counties). While a variety of statistical measures can be used to calculate the expected volume of reciprocal pairs, the following formula that is based on an adjusted flow volume was employed.

$$\mathrm{EP}\,(\mathrm{O},\mathrm{D}) = \mathrm{F_O F_D f}\,(\mathrm{O},\mathrm{D}) \,/\left(\mathrm{F}_s^2 - \sum\nolimits_{i=0}^{n} F_i^2\right)$$

where EP (O, D) is the expected number of reciprocal pairs between origin O and destination D, $\mathrm{F_O}$ is the number of reciprocal pairs between county O and its connections, $\mathrm{F_D}$ is the number of reciprocal pairs between county D and its connections, f (O, D) is the number of reciprocal pairs between county O and county D, $\mathrm{F_S}$ is the number of reciprocal pairs between all counties, and $\sum_{i=0}^{n} {F_i}^2$ is used to remove within-county expectations. Finally, modularity of a link O-D is calculated as:

$$\mathrm{MOD}\,(\mathrm{O},\mathrm{D}) = \mathrm{AP} - \mathrm{EP}$$

where AP is actual number of pairs, and EP is expected number of pairs on link O–D. Using this formula, the raw counts of reciprocal pairs were transformed into a county-to-county modularity graph, in which the weight of a link represents the modularity between two counties. If modularity value is positive the link is considered to be above expectation, if the value is negative the link is below expectation. Next, a full-order average linkage algorithm (ALK) [40] was employed to construct a set of spatially contiguous regions. One can find the algorithmic details of the clustering method in [40]. The average linkage algorithm is a clustering method which is used to build a hierarchy of spatially contiguous clusters by iteratively merging the most connected adjacent clusters. The method outputs a spatially contiguous tree, where each edge connects two geographic neighbors and the entire tree is consistent with the cluster hierarchy. Next, each region in the spatially contiguous tree was partitioned into two regions based on an objective function. Partitioning starts downward from the top of the clustering tree by removing edges. To obtain k regions, (k−1) edges must be removed. For example, four edges must be removed from the initial spatially contiguous tree to derive a five-region partition. To derive k regions, a hierarch of k sets of region partitions are obtained. Each of these sets corresponds to a hierarchical level and is embedded in the next higher level of region partition. Given two regions generated at each level of the hierarchy, a fine-tuning procedure [9] was performed to modify the boundaries by moving locations from one region to other to further optimize the objectives. In this paper, two objectives were used: (1) maximizing within-region modularity (2) maximizing compactness for each region. The modularity is the sum of flow-expectation difference for each pair of units inside a region and for all regions. Different from the original algorithm [9], we used hierarchical expectation by recalculating the marginal flows for the new region division after each edge removal. For example, if an edge removal partitions ten spatial objects into two regions, region A with three and region B with seven; the marginal flows of the three locations in A is recalculated as the marginal flows within A, and the same applies to region B. Therefore, the marginal flows and flow totals of locations in both regions are dynamically updated according to which region they belong to [41]. The compactness of a region was calculated using the Relative Distance Variance [42, 43], which was found to outperform the other measures of compactness [44]:

$$Compactness = \sqrt{\frac{Area}{2\pi\left(\sigma_x^2 + \sigma_y^2\right)}}$$

where Area is the area of the shape, and σ_x^2 and σ_y^2 represent the variance of the distances between the centroid of the shape, and the x and y coordinate pairs that define the boundary of the shape.

5 Results

5.1 Network Characteristics and the Distance Effect

Individual-to-individual reciprocal mention network consisted of 1,539,396 users (nodes) who participated in at least one conversation. There were 2,621,831 undirected edges, where each edge illustrates a reciprocal pair of users who communicated with each other at least once. Despite the extensive filtering process, the reciprocal communication network is still well connected [45]. The largest connected component consisted of 1,271,530 users (83%) and 2,424,224 edges (92%). This means that 83% of the individuals are connected with each other by a varying number of steps, and an individual has 1.9 connections on average. Figure 1 illustrates the cumulative density of reciprocal pairs by geographic distance. While 50% of the reciprocal communication happened within the same county, 77% happened within the same state. This finding agrees with the previous work in that individuals who engage in conversations are strongly constrained by geographic space.

5.2 Multi-scale Community Structures

The individual-to-individual network was aggregated into to county-to-county network of reciprocal communication, and the regionalization algorithm was performed to derive a hierarchy of regions from 1 to 48. The partition with 48 regions

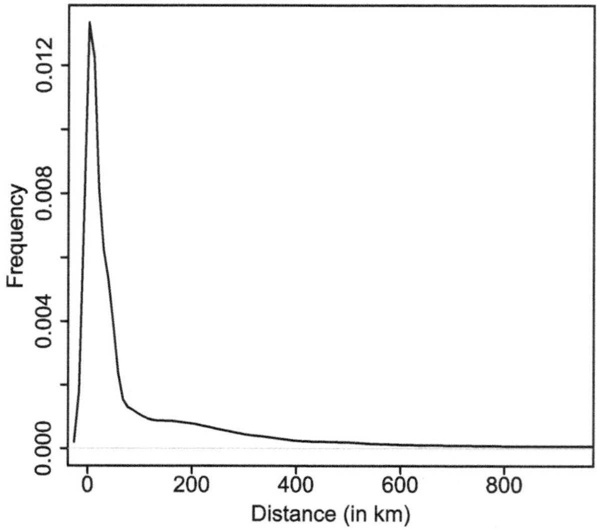

Fig. 1 Frequency of geographical distances among reciprocal pairs. While 50% of the reciprocal pairs were within the same county, 77% were within the same state

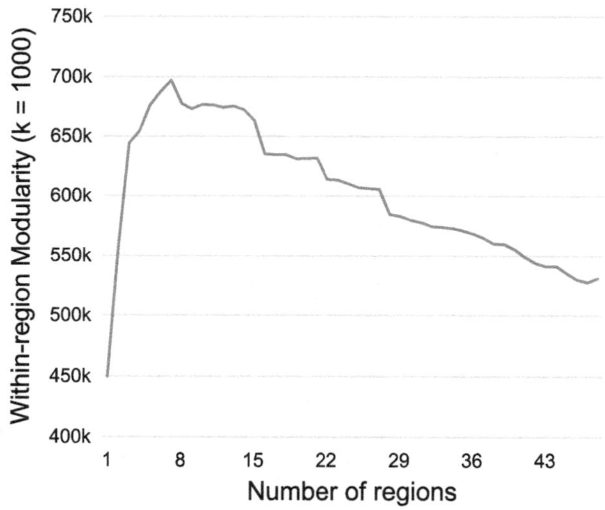

Fig. 2 Total within-region modularity for partitions from 1 to 48 regions in the hierarchy

was selected as the maximum number of regions in the hierarchy in order to compare the data-driven regions to the boundaries of the lower 48 states. The total within-region modularity for region levels from 1 region to 48 regions highlights patterns of communication at multiple scales (Fig. 2). The three-region partition (Fig. 3a) splits the country into East, Central South and Midwest-West divisions. The existence of the eastern region is likely to be influenced by different time zones, which enforce a significant constraint in human communication. The partition with eight regions maximizes the total within-region modularity, and suggests a stable partitioning

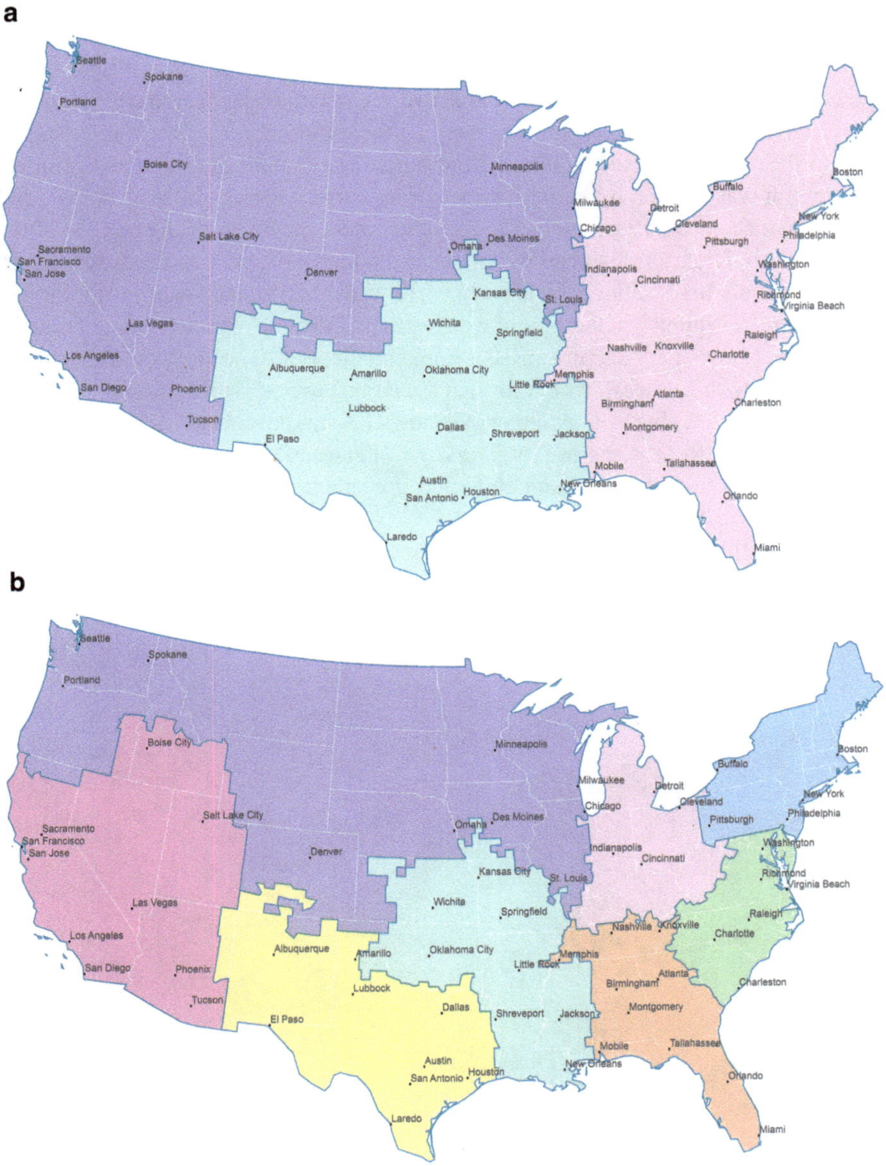

Fig. 3 Hierarchy of interpersonal communication at (**a**) three regions (**b**) eight regions. Partition with eight regions achieves the maximum within-region modularity, and suggests a stable partitioning of the network for the discovery of community structures

of the network for the discovery of community structures (Fig. 3b). Eight-region partition highlights known boundaries as well as unexpected splits that can be explained by socio-economic, cultural and dialectal, and topographical structure of the country. The Northeast region almost exactly matches the designated region by the Census Bureau. This is not surprising as the cultural and political make-up of the Northeast was established long before other regions, and over several centuries. The region was formed by various ethnic groups that were spatially clustered, and tightly connected with each other. On the other hand, the neighboring regions of the Northeast are largely influenced by the natural boundaries such as the Appalachian Mountains and Ohio Valley which act like a physical barrier, and catalyst for human connectivity. Regions in the south were split by the state boundaries of Texas, Tennessee, Louisiana, Alabama, Mississippi and Georgia. The Northwestern region was merged with Midwest, which formed the largest region with a minimal effect of state boundaries. California, Arizona, Nevada, Utah and South of Idaho formed the Western region. Regardless of the diversity in landscape and climate, the Western region contains various racial and ethnic groups that are connected with each other across longer distances.

Figure 4 illustrates 27 regions which were selected based on the most significant drop (slope) in total within-region modularity around the mid-level regions (Fig. 2). This partition highlights previously known splits in regional geography of the U.S. and patches created by metropolitan areas such as Dallas, Los Angeles,

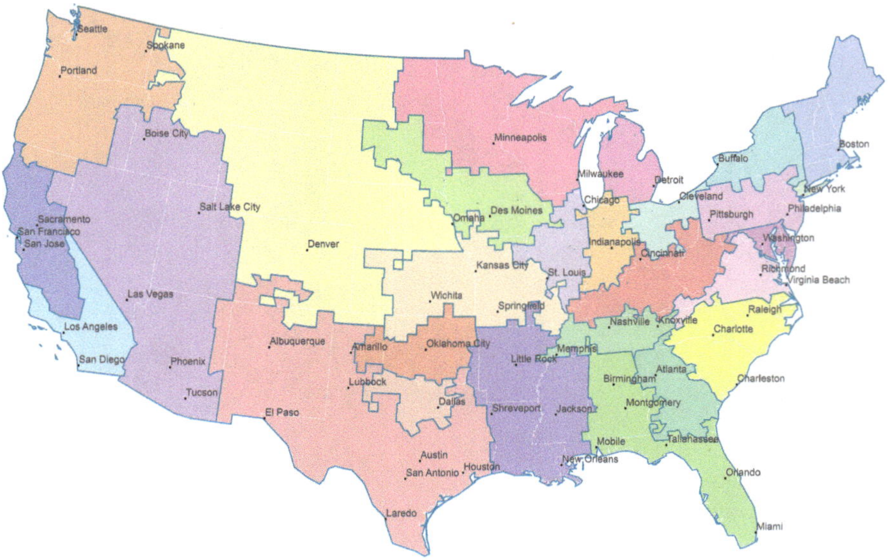

Fig. 4 Interpersonal communication at 27 regions. This partition highlights previously known splits in regional geography of the U.S. such as the division between northern and southern California; Carolinas; Great Lakes region including Minnesota, Wisconsin and Michigan; and patches created by metropolitan areas such as Dallas, Los Angeles, and Washington D.C

and Washington D.C. There are many known splits in this partition such as the division between northern and southern California; Carolinas; Great Lakes region including Minnesota, Wisconsin and Michigan; the combined Kansas-Missouri region centered on the two Kansas Cities and Springfield, Missouri; and the separation of New York City from the rest of New York.

Figure 5a illustrates the partition with 48 regions in order to compare with the boundaries of the lower 48 states of the U.S. While the regions in the east are partitioned into smaller regions, regions in the west are still very large due to lower population, thus, communication sparsity. Figure 5b illustrates the overlap between the state borders and the boundaries of the 48 data-driven regions. The overlap between the state boundaries and 48 data-driven regions was found to be 45%. The states with the most overlap with the region boundaries are Pennsylvania (83%), New Jersey (80%), South Carolina (80%) and Arizona (78%) (Fig. 5b). While some states were split into smaller regions, some were merged to form larger regions that contain multiple states. For example, Texas was split into three regions influenced by the metropolitan cores of Houston, San Antonio, and Dallas. California was split into San Francisco, Central Valley and the rest of California that is pulled by Los Angeles. Florida was split into two regions as a result of the pull effect of the metropolitan areas of Miami, and Northern Florida (i.e., Orlando, and Jacksonville). Small deviations from state borders are caused by the swapping of counties as a result of the pull-effect of a metropolitan core in an adjacent state. Some states were merged to form larger regions that include multiple states. Most of these examples are from the Great Plains. A common characteristic of these regions is the low population density, and thus, less volume of communication.

5.3 *Spatial Connectivity Between Regions*

Figure 6 illustrates the patterns of spatial connectivity between 48 regions. A modularity threshold of 500 was used to reduce the cluttering and visualize flows that are above expectation (i.e., observed—expected >500). A circle symbol is placed at the population-weighted centroid of a region and the size of the circle is proportional to the within-region modularity. Modularity flows between the regions are represented by flow lines with varying width proportional to the modularity value. Background choropleth map illustrates the region boundaries, and the color value is used to symbolize the density of reciprocal pairs within each region using quantile classification. The structure of flows follow a variety of forms. For example, the Texas Triangle portrays a polycentric pattern, where there are approximately equal strength of connections (flows) between the three metropolitan regions of Houston, San Antonio, and Dallas. On the other hand, connections in California follow a more hierarchical structure, where the hinterland of Los Angeles is tightly connected with the hubs of Central Valley, San Francisco, and Arizona; the connections between these hubs are not as strong. The regions in the East Coast, on the other hand, follow a linear pattern similar to a spanning tree, where each of the

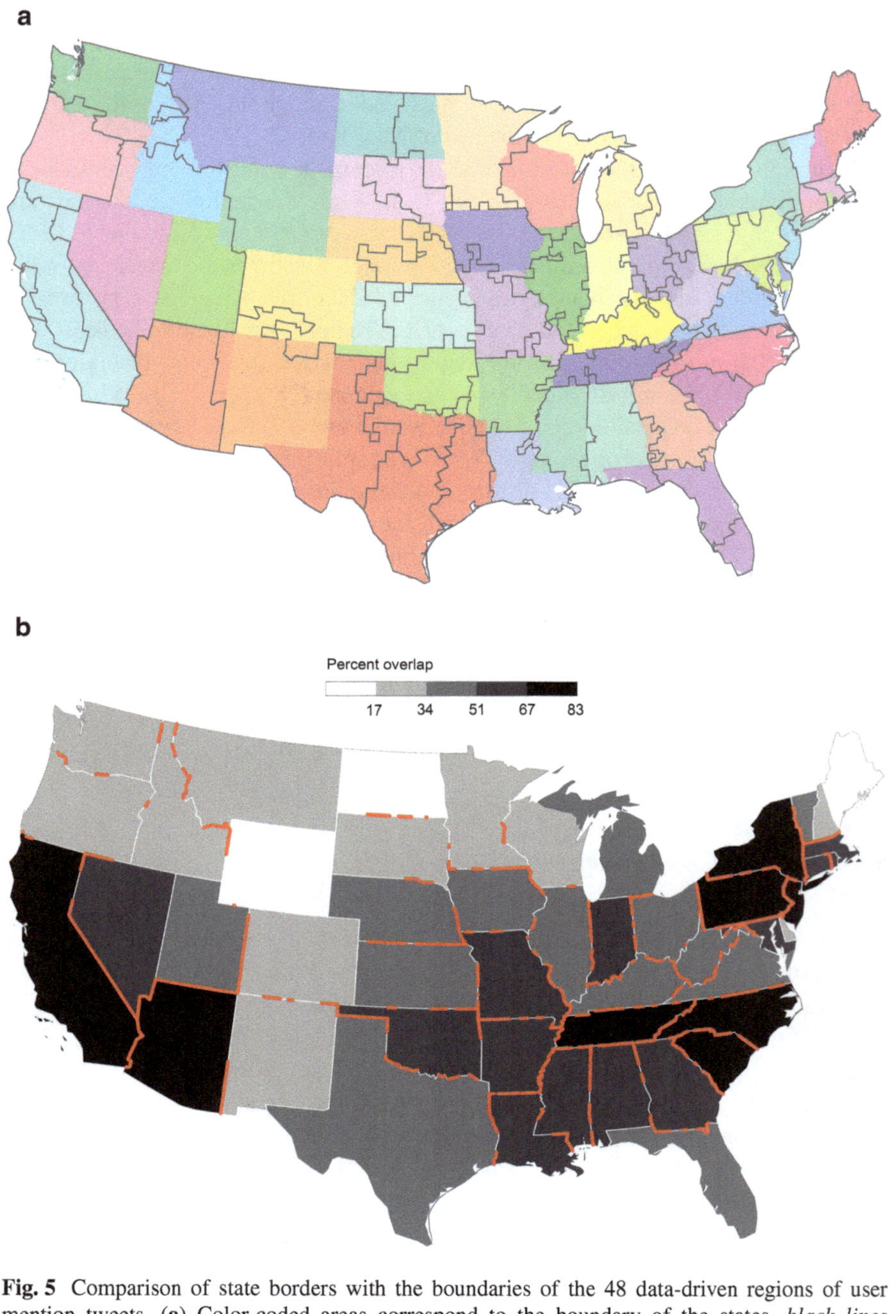

Fig. 5 Comparison of state borders with the boundaries of the 48 data-driven regions of user mention tweets. (**a**) Color-coded areas correspond to the boundary of the states, *black lines* correspond to the boundaries of data-driven regions discovered by the regionalization algorithm. (**b**) *Red lines* illustrate the overlap between the state boundaries and the 48 regions, and the color value symbolizes the percentage of overlap for each state. The overlap between the state boundaries and 48 data-driven regions was found to be 45%

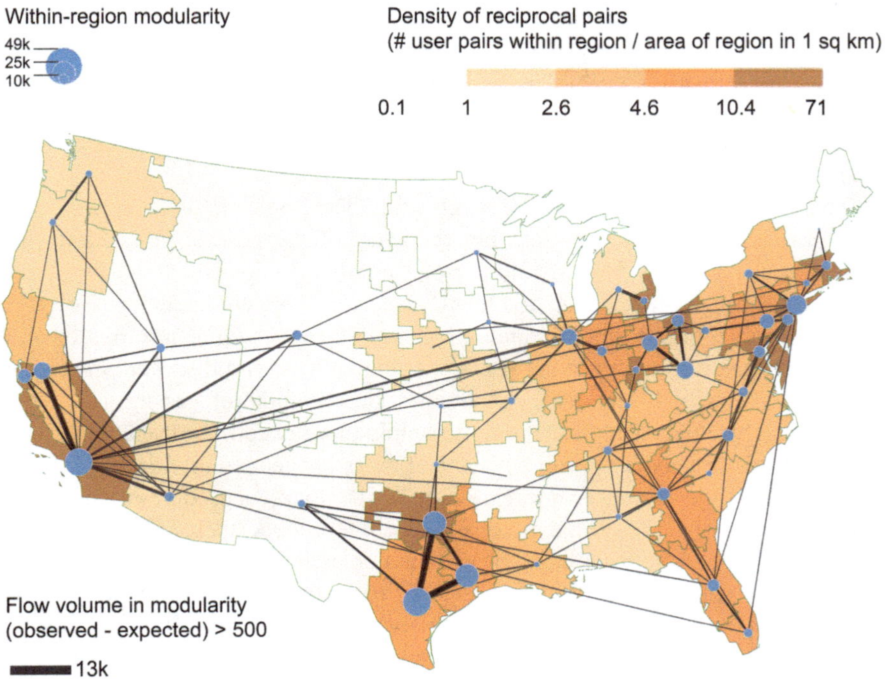

Fig. 6 Reciprocal mentions between 48 regions. A *circle symbol* is placed at the population-weighted centroid based on the number of users within each region, and the size of the circle is proportional to within-region modularity. Modularity flows between the regions are represented by flow lines with varying width proportional to the modularity value. Background choropleth map illustrates the region boundaries, and the color value is used to symbolize the density of reciprocal pairs within each region using quantile classification

regions are strongly connected to one of its close-by neighbors along the east coast. The only exception to this pattern are the big hubs of New York City and New Jersey, which follow a hierarchical pattern. Chicago also follows a hierarchical pattern of connectivity, whereas Cleveland, Columbus and West Virginia follow a polycentric one with strong connections among each other.

6 Discussion and Conclusion

A hierarchical regionalization algorithm was used to identify multi-scale community structures within the interpersonal communication network on Twitter. The results strikingly showed cohesive regions in different scales, which overlap with administrative boundaries of the states, cultural and dialectal regions, and topographical features. Although the regionalization process did not involve state level information, 45% of the state borders overlapped with the data-driven regions, which is similar to the findings of the previous studies that analyzed a variety of human mobility and communication datasets [9, 13]. Also, the patterns of spatial connectivity between the 48 regions revealed a variety of structural patterns such as poly-centricity, hierarchies, and spanning trees. Discovery of such patterns is essential for understanding of the complex social system that is influenced by long-distance ties.

There are a number of limitations in this study. The first limitation is well-known: demographics of twitter users are not reflective of the general population [46]. Twitter is only a small portion of interpersonal communication which mostly happen in person, through phone calls, text messaging, and video conferencing. However, one can analyze any form of communication data with spatial information in a similar manner without revealing privacy of individuals, and discover community structures in a spatial hierarchy. Although a large volume of geo-located tweets were used, these tweets represent only a sample of all tweets (approximately 1%). Moreover, constrained by opt-in behavior of users for geographic location, a large portion of user mentions was not represented in the datasets used in this study due to the inability to locate all mention pairs. For future work, there is a need to take into account the changing frequency of communication over time. In addition to studying the temporal aspect of the network, there is also a need to examine the semantics of the communication using the content of the tweets. By analyzing the content of the conversations using text mining methods one can understand how online conversations vary based on pairs of users in different locations, and different time periods. Such information can help identify both linguistic and topical variation across regions, and improve our understanding of complex semantics in human communication.

References

1. Backstrom L, Sun E, Marlow C (2010) Find me if you can: improving geographical prediction with social and spatial proximity. In: Proceedings of the 19th international conference on World wide web. pp. 61–70. ACM
2. Mok D, Wellman B, Carrasco J (2010) Does distance matter in the age of the Internet? Urban Stud 47:2747–2783
3. Garcia-Gavilanes R, Mejova Y, Quercia D (2014) Twitter ain't without frontiers: economic, social, and cultural boundaries in international communication. In: Proceedings of the 17th ACM conference on Computer supported cooperative work & social computing, p. 1511–1522. ACM, Baltimore, Maryland, USA
4. Takhteyev Y, Gruzd A, Wellman B (2012) Geography of Twitter networks. Soc Networks 34:73–81
5. Yardi S, Boyd D (2010) Tweeting from the Town Square: Measuring geographic local networks. In: ICWSM
6. Leskovec J, Horvitz E (2014) Geospatial structure of a planetary-scale social network. IEEE Trans Comput Soc Syst 1:156–163
7. Park P, Weber I, Mejova Y, Macy M (2013) The mesh of civilizations and international email flows. In: WebSci 2013 Proceedings. ACM

8. Kylasa SB, Kollias G, Grama A Social ties and checkin sites: connections and latent structures in location based social networks. In: Proceedings of the 2015 IEEE/ACM international conference on advances in social networks analysis and mining 2015, pp. 194–201. ACM
9. Guo D (2009) Flow mapping and multivariate visualization of large spatial interaction data. IEEE Trans Visual Comp Grap 15:1041–1048
10. Sobolevsky S, Szell M, Campari R, Couronné T, Smoreda Z, Ratti C (2013) Delineating geographical regions with networks of human interactions in an extensive set of countries. PLoS One 8:e81707
11. Chen Y, Xu J, Xu MZ (2015) Finding community structure in spatially constrained complex networks. Int J Geogr Inf Sci 29:889–911
12. Ratti C, Sobolevsky S, Calabrese F, Andris C, Reades J, Martino M, Claxton R, Strogatz SH (2010) Redrawing the map of Great Britain from a network of human interactions. PLoS One 5:e14248
13. Thiemann C, Theis F, Grady D, Brune R, Brockmann D (2010) The structure of borders in a small world. PLoS One 5:e15422
14. Slater PB (1975) Hierarchical regionalization of RSFSR administrative units using 1966-69 migration data. Soviet Geog Rev. Transl 16:453–465
15. Grauwin S, Szell M, Sobolevsky S, Hövel P, Simini F, Vanhoof M, Smoreda Z, Barabási A-L, Ratti C (2017) Identifying and modeling the structural discontinuities of human interactions. Sci Rep 7:46677
16. Deville P, Song CM, Eagle N, Blondel VD, Barabasi AL, Wang DS (2016) Scaling identity connects human mobility and social interactions. Proc Natl Acad Sci U S A 113:7047–7052
17. Emmerich T, Bunde A, Havlin S, Li G, Li D (2013) Complex networks embedded in space: dimension and scaling relations between mass, topological distance, and Euclidean distance. Phys Rev. E 87:032802
18. von Landesberger T, Brodkorb F, Roskosch P, Andrienko N, Andrienko G, Kerren A (2016) Mobilitygraphs: visual analysis of mass mobility dynamics via spatio-temporal graphs and clustering. IEEE Trans Vis Comput Graph 22:11–20
19. McGee J, Caverlee JA, Cheng Z (2011) A geographic study of tie strength in social media. In: Proceedings of the 20th ACM international conference on Information and knowledge management, pp. 2333–2336. ACM
20. Krings G, Calabrese F, Ratti C, Blondel VD (2009) Urban gravity: a model for inter-city telecommunication flows. J Stat Mech Theory Exp 2009(07):L07003
21. Lambiotte R, Blondel VD, De Kerchove C, Huens E, Prieur C, Smoreda Z, Van Dooren P (2008) Geographical dispersal of mobile communication networks. Phys A Statis Mech Appl 387:5317–5325
22. Barnett I, Khanna T, Onnela J-P (2016) Social and spatial clustering of people at humanity's largest gathering. PLoS One 11:e0156794
23. Garcia-Gavilanes R, Quercia D, Jaimes A (2013) Cultural dimensions in twitter: time, individualism and power. In: International AAAI conference on weblogs and social media
24. Yamaguchi Y, Amagasa T, Kitagawa H (2013) Landmark-based user location inference in social media. In: Proceedings of the first ACM conference on Online social networks, pp. 223–234. ACM
25. Jurgens D (2013) That's what friends are for: inferring location in online social media platforms based on social relationships. ICWSM 13:273–282
26. Compton R, Jurgens D, Allen D (2014) Geotagging one hundred million twitter accounts with total variation minimization. In: 2014 IEEE international conference on big data (big data), pp. 393–401. IEEE
27. HerdaĞdelen A, Zuo W, Gard-Murray A, Bar-Yam Y (2013) An exploration of social identity: the geography and politics of news-sharing communities in twitter. Complexity 19:10–20
28. Groh G, Straub F, Eicher J, Grob D (2014) Geographic aspects of tie strength and value of information in social networking. p. 1–10. ACM
29. Girvan M, Newman ME (2002) Community structure in social and biological networks. Proc Natl Acad Sci 99:7821–7826

30. Clauset A, Newman MEJ, Moore C (2004) Finding community structure in very large networks. Phys Rev. E 70:066111
31. Hawelka B, Sitko I, Beinat E, Sobolevsky S, Kazakopoulos P, Ratti C (2014) Geo-located Twitter as proxy for global mobility patterns. Cartogr Geogr Inf Sci 41:260–271
32. Nelson GD, Rae A (2016) An economic geography of the United States: from commutes to megaregions. PLoS One 11:e0166083
33. Kallus Z, Barankai N, Szule J, Vattay G (2015) Spatial fingerprints of community structure in human interaction network for an extensive set of large-scale regions. PLoS One 10:e0126713
34. Wang F, Mack EA, Maciewjewski R (2017) Analyzing entrepreneurial social networks with big data. Ann Am Assoc Geog 107:130–150
35. Sobolevsky S, Sitko I, des Combes RT, Hawelka B, Arias JM, Ratti C (2014) Money on the move: big data of bank card transactions as the new proxy for human mobility patterns and regional delineation. In: The case of residents and foreign visitors in Spain. 2014 IEEE international congress on big data (bigdata congress), pp. 136–143
36. Croitoru A, Wayant N, Crooks A, Radzikowski J, Stefanidis A (2015) Linking cyber and physical spaces through community detection and clustering in social media feeds. Comput Environ Urban Syst 53:47–64
37. Gao S, Liu Y, Wang Y, Ma X (2013) Discovering spatial interaction communities from mobile phone data. Trans GIS 17:463–481
38. Stefanidis A, Cotnoir A, Croitoru A, Crooks A, Rice M, Radzikowski J (2013) Demarcating new boundaries: mapping virtual polycentric communities through social media content. Cartogr Geogr Inf Sci 40:116–129
39. Lansley G, Longley PA (2016) The geography of Twitter topics in London. Comput Environ Urban Syst 58:85–96
40. Guo D (2008) Regionalization with dynamically constrained agglomerative clustering and partitioning (REDCAP). Int J Geogr Inf Sci 22:801–823
41. Newman ME, Girvan M (2004) Finding and evaluating community structure in networks. Phys Rev. E 69:026113
42. Bachi R (1973) Geostatistical analysis of territories. Bull Int Stat Ins 45:121–133
43. Blair D, Biss T (1967) The measurement of shape in geography: an appraisal of methods and techniques. Bulletin of Quantitative Data for Geographers. p 45
44. MacEachren AM (1985) Compactness of geographic shape: comparison and evaluation of measures. Geografiska Ann Ser B Human Geogr 67:53
45. Cogan P, Andrews M, Bradonjic M, Kennedy WS, Sala A, Tucci G Reconstruction and analysis of twitter conversation graphs. In: Proceedings of the First ACM international workshop on hot topics on interdisciplinary social networks research, pp. 25–31. ACM
46. Pavalanathan, U., Eisenstein, J. (2015) Confounds and consequences in geotagged twitter data. arXiv:1506.02275